# ROBERT HOOKE'S CONTRIBUTIONS TO MECHANICS

# ROBERT HOOKE'S CONTRIBUTIONS TO MECHANICS

## A STUDY IN SEVENTEENTH CENTURY NATURAL PHILOSOPHY

*by*

F. F. CENTORE, PH. D.

Springer-Science+Business Media, B.V. 1970

ISBN 978-94-017-5076-9 ISBN 978-94-017-5074-5 (eBook)
DOI 10.1007/978-94-017-5074-5


Originally published by Martinus Nijhoff, The Hague, Netherlands in 1970.

DEDICATION

*To*
*My Parents*
*Mr. and Mrs. Paul Centore*

# PREFACE

In the history of science and philosophy and the philosophy of nature the name Robert Hooke has been largely ignored. If he is occasionally mentioned, it is usually in one of two ways: either he is briefly referred to in passing, or, he is viewed through the eyes of some later giant in the history of science and philosophy such as Sir Isaac Newton. Both approaches, however, do Hooke an injustice.

In the academic world of today, there is no scholarly study available of Hooke's actual place in the history of science and philosophy with respect to his doctrines and accomplishments within the area of mechanics. Such a situation constitutes an unfortunate lacuna in the academic life of the world in our time. It is the more unfortunate because, in his time, Robert Hooke played an important role in the intellectual life of his world.

Hooke, a contemporary of Boyle and Newton, lived from 1635 to 1703. For most of his active intellectual life he held the position of Curator of Experiments to the Royal Society of London. As a result of his own initiative and of directives given him by other members of the Society, Hooke performed hundreds of experiments designed to explore the secrets of nature so that men might better understand God's creation. In this treatise I will disengage from the large disorganized welter of monographs and treatises left by Hooke all the material pertinent to the science of mechanics.

Fortunately, the vast majority of Hooke's writings on all subjects have been published in various forms so that the original sources are available for my use. It will be my task to gather and analyze all the pertinent information on this one aspect, mentioned above, of Hooke's variegated accomplishments in the hope that a fruitful synthesis of his work might be manifested. It is not my purpose to defend any one thesis (although I do reach some general conclusions) or to produce an historical biography relating Hooke's personal development.

Of course this must all be done within the context of Hooke's own time and circumstances. This means taking into consideration what influences the major earlier thinkers had upon Hooke's own views as well as how Hooke may have influenced those who came after him.

Our purpose in this treatise, then, is to gain an appreciation of Robert Hooke, to see him in perspective. This means, within the restricted scope of this treatise, attempting to see with respect to the science of mechanics, as far as is possible, the world as Hooke saw it. To this end we have composed the chapters that are to follow.

Chapter I is devoted to Hooke's life and the general atmosphere in which he worked. Chapter II becomes more detailed and gives us the foundation for our appreciation of Hooke. In it we show what Hooke both theoretically and actually thought of Francis Bacon and the Baconian philosophy of scientific methodology. In the light of Chapter II everything that follows will become to a great extent intelligible. Chapter III takes up what appears to have been, by way of Bacon and Boyle, Hooke's first love, the study of the mechanics of fluid bodies, especially the air. Chapter IV takes us into Hooke's mechanics of moving bodies on or near the surface of the earth. At this point we introduce the world view of Descartes which is needed to render Hooke's work completely intelligible. Descartes had proposed a view of nature which fitted Hooke's needs like a glove. In Chapter V we consider the motions of celestial bodies. We also examine, by means of comparison and contrast, Hooke's relationship to the Newtonian synthesis in mechanics. Finally, in Chapter VI we state what has been uncovered by our study; namely, that, on the negative side, Hooke the Baconian was greatly curtailed in his work on mechanics because he refused to give up his role as a philosopher of nature, while, on the positive side, Hooke must be revered as a great contributor to the development of science in general by his constant insistence upon the importance of experimentation.

Let me take this opportunity to say that any faults the reader may find in this work are to be assigned to me, while what is worthwhile is to be credited directly to my colleague, Dr. John J. Coffey, professor of philosophy in St. John's University (New York). To him I express my gratitude for introducing me to Hooke, indicating the need for this study, guiding my research, and encouraging me to go on when I tended to become discouraged over some aspect of this study. For their valuable comments, Fr. P. Farrell, O. P. and Dr. A. Schrynemakers, both of the philosophy department, and Dr. E. Butkov, of the physics department, have my gratitude. My wife, Helen, is also deserving of thanks for her help in typing and proofreading.

University of Waterloo
Canada, 1970

## TABLE OF CONTENTS

## LIST OF DIAGRAMS

# LIST OF ABBREVIATIONS USED IN FOOTNOTES

A. L.; N. O.: Bacon, Francis, *The Advancement of Learning and Novum Organum* (tr. and ed. by J. E. Creighton), Revised ed., New York, 1900.

Boyle: Boyle, Robert, *The Works of the Honourable Robert Boyle* (ed. by Thomas Birch), 6 vols., London, 1772.

Gunther: Gunther, R. W. T., *Early Science in Oxford*, 14 vols., Oxford, England, 1920-1945.

Herivel: Herivel, John, *The Background to Newton's Principia*, Oxford, England, 1965.

M.: Hooke, Robert, *Micrographia*, London, 1665. (Dover facsimile reproduction, New York, 1961.)

More: More, Louis T., *Isaac Newton*, New York, 1962.

P. P.: Descartes, René, *Principia Philosophiae*, Amsterdam, 1644. (As contained in *René Descartes: Philosophical Writings*, tr. and ed. by E. Anscombe and P. T. Geach, Edinburgh, 1964.)

P. W.: Hooke, Robert, *The Posthumous Works of Robert Hooke* (ed. by R. Waller), London, 1705.

Sabra: Sabra, A. I., *Theories of Light From Descartes to Newton*, London, 1967.

# CHAPTER I

# HOOKE'S LIFE AND TIMES

## *1.1 Hooke's Early Life.*

Hooke's life strikes one as monolithic in character. In each portion of the whole his interests and attitudes remained basically the same, even though he might have concentrated for short periods on one particular subject or another. If one is to discern any pattern in the Curator's career it would be the following: its rise from 1648 to 1662; its plateau period from 1663 to 1687; its denouement from 1688 to the close. He was consistently conscientious in the prosecution of his self-appointed life-long task of prying into nature's secrets and improving man's mechanical advantage over his environment.

Robert Hooke entered the world at 12 noon on Saturday 18 July 1635 according to the Old Style English Calendar.[1] He was born on the Isle

[1] The material relating to Hooke's life is taken largely from Richard Waller's Introduction to his edition of *The Posthumous Works of Robert Hooke* (London, 1705). This is the earliest and most authoritative account, although it has been augmented later by some references in John Ward's *Lives of the Professors of Gresham College* (London, 1740) and John Aubrey's *Brief Lives,* first edited by Andrew Clark and published at Oxford in 1898.

Although no picture remains, John Aubrey, Hooke's close friend, tells us that Hooke was somewhat deformed in stature, with a pale complexion and a receding chin. He possessed a relatively large head exhibiting grey, protruding eyes, and a large amount of curly brown hair usually uncut and uncombed. In addition, Hooke's memory was supposedly relatively poor, while his ability in arithmetic was inferior to that in geometry. His eating and sleeping habits were poor. If one came by at two or three in the morning, Hooke could be found still up and about working on something or other. Nevertheless, Aubrey is certain that Hooke was a man of great virtue and goodness. In contrast, Waller describes Hooke as a person who was melancholy, mistrustful and jealous. The discrepancy is easily reconciled once it is known that Aubrey, writing about 1680, speaks of Hooke as he was in his better days while Waller, remembering the man after his death, thought of him as he was near the end of his life. Furthermore, Waller adds that Hooke appeared to have a deep religious faith all his life besides cultivating an interest in Holy Scripture of no mean proof Wight, the son of the curate of Freshwater. As he tells us in an autobiography, which he began to write in 1697 but never completed, he was a sickly child whose parents were sure that he would not survive for long. He was given, in fact, no hard foods during the first seven years of life. Due to his son's weak condition, his father early despaired of ever fitting him into the scholar's life. Besides, the boy seemed much more interested in drawing and making mechanical models, toys and the like, rather than in studying his grammar.

Hooke's career began when he was only thirteen with the death of his father in October of 1648. His father left him with the sizable sum of 100 pounds plus an apprenticeship he had arranged with the painter Peter Lely. However, after his arrival in London, Hooke found that he objected to the odor of oil paints. Also, he disliked giving someone money to teach him something that he could just as well learn for himself. At about the age of fourteen he left the painter and moved into the house of Mr. Busby, the schoolmaster of Westminster. While there, Waller tells us, although we are not aware of his precise motivations, he began to avidly study Euclid, Cartesian philosophy and geometry, Latin, Greek, some Hebrew, and, in addition, learned to play the organ.

### *1.2 Hooke's Oxford Days.*

In 1653 Hooke went to Christ Church College, Oxford, as a chorister and assistant to Thomas Willis the chemist. Shortly thereafter, on Willis' recommendation, Robert Boyle accepted him as his assistant. He held this position with Boyle until late 1662. During 1655 and 1656, Hooke studied astronomy, worked on the barometer, and invented the anchor escapement mechanism for the pendulum clock. By the end of 1657 he was attempting to fulfill Boyle's desire for a more perfect air pump. The task was accomplished during 1658. Simultaneously, he endeavored to design instruments employing springs, both straight and spiral, instead of the force of gravity to keep a clock in constant uniform motion. Derham, it might be interjected, in his *The Artificial Clock-Maker* (3rd ed., London, 1714), claimed to have seen an inscription upon a spring-driven watch given to Charles II which read: "ROBERT HOOKE INVEN. 1658, T. TOMPION FECIT 1675."

The problem of finding longitude at sea was a major one in Hooke's

portions. In accordance with the rules for a Gresham College professorship, Hooke never married. See M. 'Espinasse, *Robert Hooke* (London, 1956) for a sympathetic treatment of his personality.

era. If such an instrument as anticipated by Hooke could truly be perfected, major problems in both navigation and map-making could be simultaneously solved in a beautifully simple manner. The issue was of such consequence that it occupied the time of most of the outstanding scientists of the century.[2] In fact, starting from the time of Galileo and going into the following century, several European governments thought the question important enough for economic reasons to warrant offering large rewards in exchange for a sure, practicable method.[3]

During the seventeenth century four basic methods were proposed, any of which, if it could have been perfected, would have been able to successfully complete the task. Three of these involved astronomical observations while the fourth utilized a dependable chronometer.[4] Among the former, the most widely used methods were the computation of lunar distances from the sun or a fixed star, and, observations of the moon's transit of the meridian. These depended upon the perfection of lunar theory and observations, things which were assiduously pursued by Rooke, Cassini, Halley, Flamsteed and Newton, and, later, Euler and Bradley.[5] Another observational method was to observe the eclipses of Jupiter's satellites. This method was first proposed by Galileo and was later developed by many of the same people mentioned in conjunction with the lunar approach. In theory, such methods were safe and sure, and appealed to those especially interested in celestial mechanics. In practice, however, such means for keeping time suffered from two great handicaps. In the first place, making the observations required a calm sea and clear skies. Secondly, the velocities of the various motions of the various heavenly bodies involved must be known with a high degree of accuracy. Neither of these conditions could be counted upon.

This situation led others, such as Hooke and Huygens, to experiment with some other ways for keeping time which they hoped would be far more dependable than those derived from astronomy. Huygens emphasized pendulum clocks, but these were found to suffer much from rough seas, changes in latitude, friction, and atmospheric conditions.[6] If another instrument could be devised, largely freed from such shortcomings, a great technological advance would have been made. Such an instrument was what we refer to today as some kind of clock or watch of the portable variety.

[2] See A. Mackay, *The Complete Navigator* (2nd ed., London, 1810).
[3] See W. Whewell, *History of the Inductive Sciences* (London, 1837), Vol. 1, p. 434.
[4] See A. Mackay, *op. cit.*, pp. 140-200.
[5] See W. Whewell, *op. cit.*, Vol. I, pp. 435ff.
[6] See C. Huygens, *Horologium Oscillatorium* (Paris, 1673).

In 1660, Boyle showed one of Hooke's prototypes for such a watch to Lord Brouncker and Sir Moray. The latter offered to patent the invention for Hooke. Hooke, in turn, almost signed a contract agreeing to perfect a clock that could be used at sea, for which he would have received considerable reward if accomplished. But he decided not to, as he tells us in a postscript to his *Helioscopes*, published in 1676,[7] because of one clause which stated that if anyone improved his invention he, and not Hooke, was to receive all the benefits therefrom. Instead, he resolved to keep any inventions of that type a secret, for which he was sharply criticized by some as a liar, until he was sure of securing justice for himself in the matter. Such favorable circumstances never did arise during Hooke's lifetime. It was not until fifteen years after 1658 that the Abbot de Hautefeuille of Orleans, France, announced the usefulness of a vibrating straight spring which he claimed to have invented. Whereas it was also about fifteen years later that Huygens published his detailed design for a clock utilizing a spiral spring. The controversy over priority which ensued was a source of pain to Hooke throughout the remainder of his life.[8]

### *1.3 Hooke and the Royal Society.*

It was also about this time that a society of thinkers centered at Gresham College in London was approaching the status of an organization that met regularly. This point was reached in 1660. It was not yet, however, a "Royal" society since its regal charter was not bestowed upon the group until several years later. One must beware, also, not to conceive of the Royal Society as a sudden phenomenon on the English scientific scene.

In 1579, Sir Thomas Gresham, a wealthy Londoner and financial adviser to Queen Elizabeth, died. According to his will, his large house on Bishopsgate Street in London and all the income from his estate were to go to his wife until her death. After her death the mansion and revenues were to be used to support seven professors, drawn from Oxford and Cambridge, in a relatively handsome manner in order that they might have the time and resources necessary to deliver scholarly public lectures in London. The seven areas to be covered were law, rhetoric, divinity, music, physics, geometry and astronomy. A short time after the death of Lady

[7] See the Bibliography for a complete list of Hooke's works.

[8] See A. Cummings, *Elements of Clock-Work* (London, 1766) for a defense of Hooke's priority in this matter and J. E. Montucla, *Histoire des Mathématiques* (Paris, 1802), Vol. 4, pp. 548ff for a contrary view.

Gresham in 1596, Gresham mansion became Gresham College and continued as a center of scientific activity until 1710 when the Royal Society moved to another building.[9]

Many of England's outstanding men of science held positions in the college. The first was Henry Briggs. He was followed by Greaves, Oughtred, Gunter, Gellibrand, Foster, Goddard and Lawrence Rooke. It was around this nucleus that the English scientific atom formed. Later, with the addition of other interested persons, the early Royal Society was formed. After an investigation of the available evidence, F. R. Johnson concluded that what ultimately became the Royal Society was the result of a "steady growth, from the very beginning of the seventeenth century, of association and collaboration among English scientists under the sponsorship of the Gresham professors of geometry and astronomy." [10]

The Society's initiation began while Hooke was still Boyle's assistant. We read in a memorandum dated November 28, 1660, how certain persons

> according to the usuall custom of most of them, mett together at Gresham Colledge to heare Mr. Wren's lecture, viz. The Lord Brouncker, Mr. Boyle, Mr. Bruce, Sir Robert Moray, Sir Paul Neile, Dr. Wilkins, Dr. Goddard, Dr. Petty, Mr. Ball, Mr. Rooke, Mr. Wren, Mr. Hill. After the lecture was ended, they did, according to the usuall manner, withdrawe for mutuall converse. Where amongst other matters that were discoursed of, something was offered about a designe of founding a Colledge for the promoting of Physico-Mathematicall Experimentall Learning. And because they had these frequent occasions of meeting with one another, it was proposed that some course might be thought of, to improve this meeting to a more regular way of debating things, and according to the manner in other countryes, where there were voluntary associations of men in academies, for the advancement of various parts of learning, so that they might doe something answerable here for the promoting of experimentall philosophy.[11]

[9] For details concerning the history of Gresham College see the Preface to Ward's *Lives of the Professors of Gresham College* (London, 1740). For an account of other groups which might be considered as precursors to the Royal Society, see F. R. Johnson, "Gresham College: Precursor of the Royal Society," *Roots of Scientific Thought* (ed. by P. P. Wiener and Noland, New York, 1957), pp. 328-353 and M. Purver, *The Royal Society: Concept and Creation* (Cambridge, Mass., 1967).

[10] *Art. cit.*, p. 353. Purver disagrees, saying that "The Royal Society was a brilliant exotic bird of passage at Gresham College, and with its departure Gresham's brief, reflected, glory vanished." (*Op. cit.*, p. 192.) Yet, if the early Society is to be *placed* at all, Gresham College is the only choice. Its *regular* meetings were there, its early leaders were there, and Hooke, the only full-time, professional scientist it could claim, was there.

[11] *The Record of the Royal Society of London* (3rd ed., London, 1912), pp. 7-8. The Rooke mentioned was Lawrence Rooke, a lecturer in astronomy within the College until 1657. At that time he became a professor of geometry at Gresham.

One must also take into account the influence upon the Society's early members of the English philosopher-politician Francis Bacon (1561 - 1626).[12] Rather than spend his time in the airy abstract realms inhabited by those who engaged exclusively in deductive reasoning, Bacon preferred to keep his feet firmly planted in the earth. In *The Advancement of Learning* (1605), which was later translated into Latin, and in his main work, *The New Organon* (1620), he complained that the universities of his day devoted too much time and energy to subjects like theology, philosophy and other abstract disciplines. Instead, if one wanted to know about nature, if one wanted to exercise his God-given position as ruler over nature, if, in short, one wanted to advance in the physical sciences, he must carefully learn and follow Bacon's new methodological tool which would reward him by its simplicity, the certainty of its results, and the practicality of its fruits. Basically, this method entailed the orderly collection of vast amounts of factual data. Mathematics was but an appendix to the substantial bodies of knowledge; a mere auxiliary to concrete inductions.[13] As Butterfield observes, the men who founded the Society were primarily motivated by one attitude of mind, namely, that experimentation was highly important to natural philosophy. The aim of the early Society was to collect, examine, explain and ultimately use to better mankind all sorts of facts about nature. This range of interest covered all of nature. However, whether examining regularities in nature, its curiosities, or even old wives' tales about nature, there was the rage for experimentation.[14]

That Hooke held Bacon in high esteem is clear from Hooke's letter to Lord Brouncker written about June, 1672: "I judge there is noe thing conduces soe much to the advancement of Philosophy as the examining of hypotheses by experiments and the enquiry into Experiments by hypotheses and I have the authority of the Incomparable Verulam to warrant me."

As far as Hooke's participation in, or contribution to, the foundation of the Royal Society is concerned, it appears certain that he was not an important figure in this respect. At best Hooke played a minor part and this

Wren replaced him as lecturer in astronomy. Also, it might be added, among the 41 persons named later in the record as original members, Hooke's name is not to be found.

[12] See W. E. Houghton, Jr., "The History of Trades: Its Relation to Seventeenth Century Thought," *Roots of Scientific Thought* (ed. by P. P. Wiener and A. Noland, New York, 1957), pp. 354-381.

[13] See A.L., III, 6; N.O., II, 8.

[14] See H. Butterfield, *The Origins of Modern Science* (New York, 1962), p. 127.

only in virtue of the fact that he was a major asset to Boyle during this period of time. As Boyle tells us in the Preface to his *A Defence of the Doctrine touching the Spring and Weight of the Air* (London, 1662), he depended heavily upon Hooke for help in formulating his work mathematically and for help in constructing and reading his experimental apparatus.

One finds that Hooke's name is conspicuously absent from all the accounts relating to both the establishment of the group as a formal society and its informal status previous to 1660. Although these accounts may disagree concerning the exact place and date of the Society's origin, they all seem to be of one mind concerning the lack of Hooke's influence. Thomas Sprat in his *History of the Royal-Society of London*, the first official account, published in London in 1667, had much to say about the importance of Oxford with respect to the Society but nothing with respect to the importance of Hooke. Likewise with John Wallis' pamphlet entitled *A Defence of the Royal Society: An Answer to the Cavils of Dr. William Holder* published in 1678. Wallis, however, defended the priority claim of Gresham against Holder who maintained that the Society had its origin at Oxford in 1649. Thomas Birch never tired of relating Hooke's accomplishments but made no attempt to credit him with any part in the foundation of the Society. And finally, mention should be made of C. R Weld's *A History of the Royal Society* (2 vols., London, 1848) the most complete account there is to date. Although Weld quoted some remarks of Hooke's criticizing the Frenchman, Cassini, for claiming it was Oldenburg, supposedly following the example set by the French, who inspired the English to found a scientific society, Weld made no statement with respect to Hooke as one of the originators. The importance of these considerations is to clearly indicate that Hooke had yet to make his place in history.

In April of 1663, however, Hooke was launched on a career that would occupy the rest of his life. He published at that time a well-received work attempting to explain observations made by Boyle, and others, on capillary action. In November of the same year, Boyle and Moray recommended Hooke as Curator of Experiments to the Royal Society. The election was unanimous. His task as Curator was to be the person who actually arranged for and performed the desired experiments. Also, 1663 saw Hooke granted an M. A. by Oxford, and in June, 1664, he began preparing to deliver a periodic series of lectures on mechanics and related topics before the Society for which he was to receive 50 pounds a year thanks to a foundation set up in that month by Sir John Cutler. Not long afterwards, Hooke was appointed a professor of geometry in Gresham College, thus permitting him to reside there, as well as curator of the Colwell collection of rarities possessed by the College.

The plague of 1665-1666 temporarily put a halt to the activities of the Royal Society. During this time Hooke lived in Surrey with Wilkins and Petty. He was, nonetheless, around long enough to judiciously observe that the cause of the plague might indeed be something carried by the air since the number of insects and other airborne creatures was greatly reduced while the plague was in process.

To add to England's troubles, there occurred in September of 1666, a great fire in London which destroyed a large part of the city. What proved to be a disaster to many, however, proved to be a boon to Hooke since it allowed him the opportunity to display his talents as an architect which otherwise might have gone completely unnoticed. Hooke, along with Oliver, was chosen as a surveyor responsible for rebuilding the city. And although his general design for the city was not accepted by the king (it is doubtful, in fact, if he ever saw Hooke's plan before accepting that of Christopher Wren's), Hooke was responsible for designing many of the buildings, among which were Bethlem Hospital, Aske's Hospital, the Physicians College, London Theatre, and Montague House (on which site now stands the British Museum), as well as the monument known as "fish-street piller." [15] Also, Hooke managed to collect a large amount of money from those willing to pay extra to have their land surveyed first and quickly. It is said he deserved the money because of his already overworked condition. Most of it was found after his death carefully stored in a large chest in his house. It was this money which formed the largest portion of his posthumous estate.[16]

Several years later, Hooke became involved in a dispute with Hevelius over whether or not plain or telescopic sights should be used on astronomical instruments; that is, whether readings taken with the naked or with the aided eye were superior. Hooke maintained the latter position while Hevelius refused to move an inch from his affirmation of the former. The debate terminated in 1671 with the publication of *Mesure de*

[15] See J. E. Elmes, *Sir Christopher Wren and His Times* (London, 1852) for an account of the important role played by Hooke as Wren's assistant.

[16] It should be noted here that the charge, made by some such as Merton, that he was primarily interested in making money is unfounded. He was primarily interested in getting the credit due him as a result of his inventions. The monetary rewards were of secondary importance. See R. K. Merton, "Science, Technology, and Society in Seventeenth Century England," *Osiris*, Vol. 4 (1938), pp. 360-632.

*la Terre* by J. Picard (1620-1682) who had employed telescopic sights with great success.[17]

Four years later, in 1674, Hooke again found himself in the middle of another debate. This time it was with Oldenburg (1615-1677), the secretary of the Royal Society from 1665 to his death, over the printing in the *Philosophical Transactions*, dated 12 March 1674, of a description of Huygens' watch without mentioning that Hooke had come upon the idea first. In the debate which followed, Oldenburg asserted that Hooke had not actually proven his claim to priority by publicly demonstrating a working example of his invention before Huygens had done so. Hooke was very much hurt and troubled over the whole issue, considering it to be a mean slight by a fellow Society member who should have taken pains to defend him rather than publicly attack him. It was this debate which led Hooke to add a "Postscript" to his *Lampas* explaining his previous silence on the subject and criticizing the actions of the secretary. It might be added that the Council of the Royal Society voted to completely disassociate itself from the postscript. Henceforth, the relationship between Hooke and Oldenburg was anything but cordial as can be seen from a letter from Hooke to Aubrey, dated 24 August 1675, in which Hooke refers to Oldenburg as a "forreine spye" and even goes so far as to suggest establishing another scientific society free from Oldenburg's influence.[18]

The following year, Hooke's secondary position as librarian and curator of collections to the Royal Society was taken over by R. Shortgrave and later by W. Perry thus allowing Hooke more time to devote to his experimental enterprises. In 1677, with the death of Oldenburg, who was responsible for publishing the Society's journal from its beginning in March of 1665 to June of 1677, Hooke thought he had sufficient time to handle the office of secretary. However, the few issues of the *Philosoph-*

[17] For background material on this debate see J. W. Olmsted, "The 'Application' of Telescopes to Astronomical Instruments 1667-1669: A Study in Historical Method," *Isis*, Vol. 40 (1949), pp. 213-225.

[18] For studies of the relationship between Hooke and Oldenburg see E. Andrade, "Robert Hooke and His Contemporaries," *Nature*, Vol. 136 (1935), pp. 358-361, and A. R. and M. B. Hall, "Why Blame Oldenburg?" *Isis*, Vol. 53 (1962), pp. 482-491.

Oldenburg was born in Bremen, Saxony, Germany. He was educated in theology at Bremen and moved to England about 1640. In 1653 he was appointed by the council of Bremen to represent it in England in matters of trade. In 1656, however, he went to Oxford where he became close friends with Boyle and Wilkins. Subsequently, he forgot about theology and economics and devoted his life to spreading the new experimental philosophy, largely through extensive correspondence and the *Philosophical Transactions* (which he published at his own expense). For what few details are known concerning Oldenburg's life, see H. Rix, "Henry Oldenburg, First Secretary of the Royal Society," *Nature*, Vol. 49 (1893-1894), pp. 9-12.

*ical Transactions* that were published between June of 1677 and the assumption of the task on a permanent basis by Robert Plot in January of 1683 were handled by Nehemiah Grew between 1678 and 1679. Nevertheless, Hooke, with the aid of others, acted as recording and corresponding secretary from October of 1677 to July of 1682. In addition, Hooke was acting as editor of a periodical entitled *Philosophical Collections* which published seven issues in London between November 1679 and April 1682. Restraining any urge he may have had for his own aggradisement, Hooke contributed only two articles to the third issue. It was also at this time that he was considering the cause of the tides and observing the famous comet of 1677.

August, 1678 found him composing a catalog of books given to the Royal Society by the Duke of Norfolk. While in December of the following year, he attempted to prove the diurnal motion of the earth by dropping a little weight from a great height and observing that it would fall to the S.S.E. of perpendicular. Earlier, in 1669, Hooke devised a zenith telescope to prove the earth's yearly rotation via direct observations of the heavens. However, according to modern standards, both attempts were unsuccessful although, referring to trials with falling bodies performed on 22 January 1680, Hooke claimed in his *Diary* to have established the "Diurnall motion of the Earth." In addition, the Curator experimented with various kinds and sizes of pendulums, some up to 200 feet long. In retrospect, one cannot help but wonder how close the world came to having an anachronistic Foucault.

The last important set of lectures delivered by Hooke transpired between April of 1681 and the end of 1682. These were concerned with the nature of light and color. It is well known that Newton delayed the publication of his *Opticks* until after Hooke's death in order not to fan into flames the embers of a new controversy emanating from questions of priority in various details of the work. Concerning another topic, we know from material preserved for us by Waller in *The Posthumous Works*, that during this same period Hooke was interested in offering a mechanical explanation for memory, and in also explicating how one arrives at the notion of time.

The general impression one receives from the study of Hooke's life is that he was a man similar to Leonardo da Vinci. He was constantly overwhelmed with new and daring ideas, sweeping schemes for the improvement of one thing or another, an imagination teeming with novel designs, and yet never quite managing to find the right circumstances required to bring into reality what was so real in his dreams. It has been said that people constantly tend to judge themselves on the basis of what they themselves believe they could do if given the opportunity, while others tend to just as consistently judge them on the basis of what they have actually accomplished. Given this premise, one can understand how a man like Hooke could have often been vexed to learn that someone else had published something which Hooke believed was rightly his because he was the first to entertain the germinal notion of the new device or technique.

It is interesting to note in this regard that Waller, throughout his whole biography, written while he was secretary and Newton president of the Society, never mentioned Hooke's debate with Newton concerning the law of universal gravitation. Perhaps this indicates its relative unimportance among Hooke's contemporaries. Then again, as will be discussed in a later chapter, the situation might not be quite so simple.

However, whereas Waller has nothing to say about Hooke's complaint before the Royal Society on 28 April 1686 (occasioned by the presentation of book I of Newton's *Principia*) that Newton was purloining the fruits of other men's labors, Ward did make the following statement: "But he seems, in some instances at least, to have carried these pretensions too far; particularly in his claim to several things in the theory of Sir Isaac Newton's *Philosophiae Naturalis Principia Mathematica,* which that illustrious writer has shown to have been his own." [19] Newton, nevertheless, did see fit to add a scholium to the fourth proposition of book I reading: "The inverse law of gravity holds in all the celestial motions, and was discovered also independently by my countrymen Wren, Hooke and Halley."

### *1.4 Hooke's Denouement.*

It is also about this time, being now close to fifty years old, that Hooke began to show signs of his advancing age, due both to his congenital weak-

[19] J. Ward, *op. cit.*, p. 188. Others find it equally hard to justify Hooke's claim. Cf. a letter from Halley to Newton dated 29 June 1686 preserved for us in D. Brewster, *Memoirs of the Life, Writings, and Discoveries of Sir Isaac Newton* (Edinburgh, 1855), Vol. I, p. 293, which reads in part: "I declared the ill success of my attempts, and Sir Christopher to encourage the inquiry, said that he would give Mr. Hooke some two months' time to bring him a convincing demonstration thereof, ... Mr. Hooke then said he had it, but that he would conceal it for some time, that others trying and failing might know to value it when he should make it public. However, I remember that Sir Christopher was little satisfied that he could do it, and though Mr. Hooke then promised to shew it him, I do not find that in that particular he has been so good as his word." For a defense of at least Hooke's non-mathematical priority see L. D. Patterson, "Hooke's Gravitation Theory and its Influence on Newton," *Isis,* Vol. 40 (1949), pp. 327-341; Vol. 41 (1950), pp. 32-45.

ness and austere way of life. His friends noted his becoming more reserved, melancholy, and cynical. After the beginning of 1687, following the death of his niece, Grace, who had lived with him for several years as a housekeeper, his condition took another turn toward greater seclusion and, by 1689, he was obviously physically declining. As his physical well-being lessened so did his experimental productivity. The record of Hooke's work from this period of his life to the end is sketchy at best.

As far as his mental capacities were concerned, however, Hooke did not appear to be suffering from any sort of dementia. He was actively engaged in suing Cutler for refusing to pay him his yearly stipends, a case which Hooke won several years later on the exact day of his sixty-first birthday. In the interim, he was given a Doctor of Physick degree in 1691, while managing to lecture on the significance of the tower of Babel in 1692, and on Ovid's *Metamorphosis* in 1693.

Beginning in July in 1697, it was clear that Hooke's sojourn on earth was drawing to a close. He seemed to have all the signs of a combination of diabetes and scurvy. His legs were swelling (they were completely black at the time of his death), intense headaches and dizzy spells sometimes caused him to fall and hurt himself, and blindness gradually overtook him. He expected to die at any moment. The end finally came on 3 March 1703 at the age of 67 years, 7 months and 13 days. His body was buried at the Church of St. Hellen in London, the whole of the Royal Society being in attendance at the funeral. Since he left no will, one cannot be sure of his last intentions.

### *1.5 Hooke's Works and Work.*

To attempt a detailed accounting of Hooke's gigantic output within the confines of this study, except with respect to his accomplishments directly related to mechanics, would be an impossible feat. To accomplish such a feat one would have to reproduce the following: William Derham's *Philosophical Experiments and Observations of the late eminent Dr. Robert Hooke, S. R. S. and geom. prof. Gresh., and other eminent virtuoso's in his time* (London, 1726); Thomas Birch's *History of the Royal Society of London* (4 vols., London, 1756); and R. W. T. Gunther's *Early Science in Oxford* published at Oxford. Gunther's 14 volume work is by far the richest general source of information on Hooke. Derham and Birch reproduced Hooke's reports to the Royal Society. Gunther collected this material from their works and reprinted it, under specific dates, in volumes 6 and 7 (1930) of his series. He also reproduced Hooke's *Cutlerian*

*Lectures* (vol. 8, 1931). In addition, Gunther has reprinted Hooke's later *Diary* (1 November 1688-8 August 1693) in volume 10 (1935). Volume 13 (1938) of the same series contains a facsimile reproduction of Hooke's *Micrographia.* Besides Gunther's volumes we have Waller's edition of *The Posthumous Works of Robert Hooke* (London, 1705) and Hooke's early *Diary* (1 August 1672-31 December 1680) only recently edited by H. W. Robinson and W. Adams (London, 1935).

Unfortunately, it should be noted further, some of Hooke's thoughts will never be known to us. Some of Hooke's papers appear to have been irrevocably lost to spite the efforts of such ardent searchers as W. Derham and Sir G. Keynes. The former, in his "To the Reader" in the work cited above, stated that he believed some of Hooke's papers were not given to him by Waller's relatives after Waller's death. Derham, an admirer of Hooke's, had hoped to complete the work started by Waller, who had originally received Hooke's papers from one of Hooke's nieces after his death, but found that he could not since some of Hooke's papers either never got to Waller, were lost by him, or, were not given to Derham but rather to other, unnamed, parties. We also read in Thomas Hearne's *Diary* (contained in *Reliquiae Herniane,* Vol. 2, Oxford, 1857, 7 April 1726) that many of Hooke's papers were lost. Hearne (1678 - 1735) was a younger contemporary of Hooke's who knew him from meetings of the Royal Society. The title of one of these lost works is known: *A Discourse of a New Instrument to Make More Accurate Observations in Astronomy.* It was written about 1661 and is known to have existed from references to it by Aubrey and Huygens. For details on the published and unpublished works of Hooke, one can do no better than to consult Geoffrey Keynes' admirable book *A Bibliography of Dr. Robert Hooke* (Oxford, 1960).

For the purposes of this study I surveyed all of the available material published by Hooke or his editors with a view toward segregating those works bearing upon the topic in hand. Also, the titles and descriptions of unpublished manuscripts, contained in Keynes' *A Bibliography of Dr. Robert Hooke,* were carefully scrutinized. As a result, the author discovered that the following works bore careful reading because they contained the heart of Hooke's work and views on mechanics: *Micrographia* (London, 1665), *Cutlerian Lectures* (London, 1679), *Posthumous Works* (London, 1705), and the reports of Hooke's work abstracted from the records of the Royal Society and their journal, the *Philosophical Transactions* (vols. 1-17), by Birch, Derham, and Gunther.

Presently, there are no thorough secondary studies on Hooke's contributions to science in general or to any particular branch of science. The best accounts, in general terms, of Hooke's achievements are Andrade's "Robert Hooke" in the *Proceedings of the Royal Society* (Series B, Vol. 137) and Margaret 'Espinasse's *Robert Hooke* (London, 1956).

Since we cannot say everything there is to say about Hooke, let it suffice for our present purposes to but briefly mention some of Hooke's major works and activities that will not be discussed in this study. The Curator was interested in ways of improving transportation, and attempted, from his earliest days in London, to design and perfect various devices for heavier than air flying. As far as this latter undertaking was concerned, he designed several heliocopter-like constructions which he was sure could be made operative provided there existed an engine capable of turning the blades fast enough. Secret codes and universal languages very much interested him. He experimented with various new uses for gun powder, and new methods for glass-blowing. With respect to measuring devices, Hooke designed instruments with which to make other tools and instruments, with much more precision than ever before, invented means for measuring the depth of the sea, air pressure, wind velocity, rain fall, and constructed a superior micrometer. He did work with acoustics, magnetism, metallurgy, geography, anatomy of muscles, and histology of plant cells. He made improvements in telescopes, burning glasses, lighting fixtures, quadrants, pendulum systems, and of great importance for the times, beginning about 1663, carried on the work started by Leeuwenhoek with the microscope. We find the major part of this work published in Hooke's *Micrographia,* one of the prides of the early Society. It must be borne in mind that for over thirty years Hooke initiated experiments, and/or carried out the experimental suggestions of others, before the Royal Society.

As a preamble to what is to follow, listen to R.W.T. Gunther as he tells us that Hooke was one of the most brilliant "men of science who took part in the really great Oxford movement, which resulted in the foundation of the Royal Society, and in establishing its prestige. All his work is characterized by pre-eminent sanity and utility." [20]

Today, even though he did not possess the mathematical genius and inclinations of a Newton, the Curator can justifiably be called a great success in the history and philosophy of science according to the canons of modern experimental research. Bacon, in his unfinished plan for a new type of social life based upon his inductive method entitled *New Atlantis,* called those people cooperating in this effort "Fellows." The term was

[20] Gunther, Vol. 6, Preface.

later adopted by the Royal Society. Bacon's fable tells of a Spanish seaman who, after drifting ashore in a distant land where the people were so advanced they appeared to be angels, is led to the "Strangers' House" where he is told the story of Atlantis. Three thousand years ago its inhabitants engaged in world trade and were known throughout the civilized world. Then came the Great Flood which nearly destroyed them. About two thousand years ago, while their society was still young, a great king, who wanted only the good of his people, came to rule them. His name was Solomon, and he passed strict immigration and travel laws in order to isolate Atlantis. But of most importance, he established Solomon's House or the "Colledge of the Sixdays Works" dedicated to the study of God's creation.

Later, one of the Fathers of Solomon's House, who spoke Spanish, condescended to tell the seaman what went on there. He summarized the End, Preparations and Instruments, Employments and Functions of the Fellows or Brethren, and the Ordinances and Rites of his organization. Its purpose was the betterment of society materially speaking, its "Riches" consisted in all varieties of experimental apparatus and situations (e.g., orchards, caves, furnaces, etc.), its members occupied themselves in gathering, arranging and disseminating information on nature that would benefit man, and its Ordinances consisted in maintaining a gallery of inventions, a gallery of inventors commemorated by statues, and daily religious services. The seaman is then sent away to tell the world of the new scientific Atlantis.

If Bacon were to return to earth today, he would find his dream largely fulfilled. The emphasis upon experimentation today in the natural sciences (along with the emphasis upon mathematics) is so obvious that it is unnecessary to discuss it. It only remains to point out that Hooke, three hundred years ago, was not only emphasizing the need for experimentation but actually practicing it. The Curator prided himself on the number and usefulness of his experiments. His thoughts, and those of Bacon's, pulsated in unison. If one could but know nature as it really is, one could twist and mold its activities into channels never before dreamed of. The experimental probings, the philosophical delvings into causes, and the arduous but necessary collecting of data, Hooke believed deep down in his heart, would some day place future man as far ahead of men in his own day as they were ahead of animals in their understanding and control of nature. All of this has, by and large, come to pass. Hooke, whose mind overflowed with inventions, experiments, and physical explanations of natural phenomena, should be highly regarded today as a pioneer in the field of experimental research.

CHAPTER II

# THE NEW EXPERIMENTAL PHILOSOPHY

## *2.1 Baconian Induction.*

Francis Bacon (1561-1626) was primarily a methodologist. In his efforts to expound a methodology suitable for the pursuit of the sciences, Bacon revealed a system which may be characterized under three heads: a desire for certainty of knowledge, a simplicity of procedure, and practical significance.

Bacon is quite clear in affirming that the results obtained via his methodology can be held with certitude. Bacon's *New Organon* starts with a "Plan of the Work," an outline of his six-part plan (called the "Great Instauration") for restoring man to his rightful place as ruler of nature, of which the *New Organon* is the second part. In Bacon's "Plan" the enquirer learns that "what the sciences stand in need of is a form of induction which shall analyze experience and take it to pieces, and by a due process of exclusion and rejection lead to an inevitable conclusion." Bacon looked upon himself as a "trumpeteer" declaring a position for his time fundamentally opposed to the logicians of the schools whom he berated for practicing a "gross" and "stupid" form of induction. Any attempt, emphasizes the Baron, "to conclude upon a bare enumeration of particulars... without instance contradictory, is a vicious conclusion." [1] The most the natural philosopher could hope for from such an enumeration would be probable conclusions. For who can say with certainty that he has taken into account all relevant particulars supporting his case and "that there are not others on the contrary side which appear not." [2] Apparently Bacon believes it is possible to reach undisputable conclusions in natural philosophy provided the *proper* method of induction is consistently employed. It is not so much that Bacon's pre-

[1] A.L., V, 2.
[2] *Loc. cit.*

decessors had failed to recognize the inductive method, as that they had failed to recognize the *proper* inductive method. In this affirmation of Bacon's resides the newness of his instrument.

But how can Bacon claim certainty as the result of any type of induction? The answer is that it must be a complete induction. But upon what material, out of all the vast array of possible experiences, should one concentrate his efforts aimed at obtaining a complete induction? Bacon's answer is that the investigator must concentrate upon analyzing simple "Forms," i.e., he must attempt to discover the inner natures of things. It is impossible, thinks Bacon, for the scientist to ever understand absolutely everything about nature. The best he can hope for is an understanding of the more fundamental, simpler aspects of nature.[3] When nature is broken up into its basic constituents, it is possible, believes the Viscount, to discover the inner constitution of the various elements considered. Rather than attempt to understand some gross object of nature, such as an animal, viewed as a whole with all its multifarious activities and attributes, the wiser scientist will instead divide up the larger whole into its more elementary aspects. It is these elementary constituents composing the natural universe that it is profitable to analyze. To use an analogy, one should approach speech from the point of view of the alphabet when endeavoring to analyze speech rather than attacking the problem from the point of view of the endless variety of sounds and combinations possible among the basic letters comprising the alphabet.

It was Bacon's sincere belief that, within a few years after the establishment of his new methodology via the agency of the King, Pope, or some group of influential individuals, man could be restored to his original place as the master of nature. As his life passed on, however, few followers arrived to carry forward his sweeping plans.[4] And, as it became more and more evident that nature was more difficult to comprehend than he had originally anticipated, he was forced to extend the time required for the great restoration to several generations. He never, nonetheless, abandoned belief in the finitude of the object to be searched out and the time required to complete the undertaking. Given the proper inductive method and sufficient facilities to put it into practice, one could in time arrive at a certain knowledge of the constitution of the universe by an exhaustive process of elimination.

Moreover, Bacon maintains that it is precisely because his methodology is certain that the conclusions reached are infallible. According to Bacon,

[3] See A.L., II.
[4] See A.L., I.

the major issue centers around the establishment of his program and not whether or not it could operate properly once established. It is in this respect that the various "Idols" mentioned by Bacon assume a great importance. It is the "Idols" which pose, as far as Bacon can see, the great hindrance to the institution of his methodology. It is these perversions of thinking that explain scientific error rather than his methodology or a fundamental inability on the part of the investigator to know with certitude. The Sophists are condemned by Bacon for having "denied that certainty could be attained at all . . . For the holders of that doctrine assert simply that nothing can be known; I also assert that not much can be known in nature by the way which is now in use. But then they go on," he continues, "to destroy the authority of the senses and understanding, whereas I proceed to devise and supply helps for the same." [5] The Baron Verulam seems to be expressing a need to clear the land, so to speak, before any new edifice can be erected when he desires the destruction of the mental impediments which he holds responsible for infecting the sciences with error. Scientific work can then be done by almost anyone.

Bacon's methodology is designed for unskilled labor. No longer must one employ the subtlety and wit of the complicated logician; one need only advance upon nature with the honesty and straight-forwardness of the common man. Others propose complicated discourse as a means of exposing nature; they anticipate nature when they should be following nature. But the course Bacon proposes for gleaning nature's secrets "is such as leaves but little to the acuteness and strength of wits, but places all wits and understandings nearly on a level. For as in the drawing of a straight line or a perfect circle," illustrates Bacon, "much depends on the steadiness and practice of the hand, if it be done by aim of hand only, but if with the aid of rule or compass, little or nothing; so is it exactly with my plan." [6] If man would master nature, he must have recourse first and foremost to naked experience.

Bacon's procedure reduces itself to the simple inspection of three tables of information. In the first table a particular phenomenon, such as heat or whiteness, is investigated as to its presence in the most diverse and varied circumstances. In the second table, that of absence, is tabulated all the instances similar to those found in the table of presence, but in which the particular phenomenon does not occur. Thirdly, a table of degrees, differences, or comparison is arranged showing the degree to which the particular phenomenon is manifested in each of the similar instances. Once

[5] N.O., I, 37.
[6] N.O., I, 61.

these three tables of appearances (the *Comparentia*) are put in order, it becomes a simple task to proceed with the method of exclusion *(Exclusiva)* which will result in the isolation of the simple nature or Form of the particular phenomenon.[7] What could be simpler? All that is required is a tabulation of appearances which any person with ordinary sensations can accomplish and some elementary reasoning to determine which characteristics are present in all the positive cases, missing in all the negative cases, and which increase when the given natural occurrence increases in frequency or intensity.

The results of the three table method, however, are only the "first vintage." They represent for Bacon the rough conclusions which must be further refined before definitive results are obtained. The refining process would be carried out by the application of what Bacon called his nine aids. Bacon never finished the second part of his *New Organon* in which these aids were to be fully explained. Only one was ever developed: the "prerogative instances." Bacon lists twenty-seven types of prerogative instances. These instances are those data or cases which are privileged; that is, which are to be given more weight than other experiences in the process of trying to reach some definitive understanding of some particular phenomenon.

Let us mention but one of the twenty-seven which proved to be of most interest to Royal Society members, especially Hooke. In the fourteenth rank of privileged data Bacon lists the decisive, judicial, or crucial instances. The crucial instances are cases in which *one* experience can make the difference between accepting one explanation of a phenomenon rather than another even if it is only a choice between a theory and its negation.

For example, one may wish to test the magnetical theory of gravitation. According to this theory the earth is a magnet which pulls bodies down to it. If this is the case, says Bacon, one would note variations in the motion of a weight-driven clock at various distances from the earth's hard core. To test the theory one should experiment with a pendulum clock far above the earth and then far within the earth and observe its slowing down or speeding up with respect to a "standard" clock on the earth's surface. He concludes that "if this power be found to diminish at a height, and to increase in subterraneous places, the attraction of the corporeal mass of the earth may be taken as the cause of weight." [8]

The general conclusion to Bacon's relatively long section on crucial ex-

[7] See N.O., II, 11-20.

[8] N.O., II, 36.

periments is a reiteration of his whole philosophy of science: "Let this suffice for the instances of the cross. We have dwelt the longer upon them in order gradually to teach and accustom mankind to judge of nature by these instances, and enlightening experiments, and not by probable reasoning." [9]

As far as the practical significance of his work is concerned, Bacon declares that all the knowledge obtained by way of his new organon is "to be referred to use and action." The whole great sweep of Bacon's written works is oriented toward the practical point of view. The vast process of data collection, the careful comparisons, the discovery of simple natures, and ultimately the finding of the Form (i.e., the true definition of the phenomenon being investigated) are all intimately related parts of a methodology which "would enfranchise the power of man unto the greatest possibility of works and effects." For, whoever knows any "Form knoweth the utmost possibility of super-inducing that nature upon any variety of nature." [10] The Forms of any phenomena could be determined by properly abstracting from concrete events and situations. After this had been accomplished the process could be reversed and the fruits of the tedious hours spent in determining the true natures of natural phenomena could be seen in a rich harvest of new and wonderful commodities to endow the life of man. This indeed would be power.

### 2.2 *Hooke and Baconian Induction in Theory.*

As with the Royal Society in general, Hooke was greatly influenced by Francis Bacon. The influence of Bacon on the Society and Hooke has already been noted in several places. It now remains for us to see in more detail the relationship between Bacon and Hooke.

The most concise theoretical statement of the spirit of Baconian methodology as understood by Hooke is to be found in his "A General Scheme, or Idea of the Present State of Natural Philosophy, and How its Defects may be Remedied by a Methodical Proceeding in the making of Experiments and collecting Observations. Whereby To Compile a Natural History, as the Solid Basis for the Superstructure of True Philosophy." This treatise (undated) was probably written about 1667 but did not appear in print until Waller fittingly published it as the first paper in *The Posthumous Works of Robert Hooke.*

Hooke begins his treatise by telling his readers that the aim of experi-

[9] *Loc. cit.*
[10] A.L., III, 4.

mental philosophy should be the discovery of the nature and properties of bodies as well as the true causes of natural phenomena. In a Baconian manner Hooke goes on to insist that such knowledge is not to be gleaned purely for its own sake. Instead, all knowledge was to be directed toward improving the material well-being of mankind. It was Hooke's hope that the men of the future would be to present men as present men are to "Brutes or Ideots." [11]

After outlining, in a fashion closely paralleling Bacon's "idols," the reasons why philosophy has not prospered in the recent past, Hooke goes on to strongly affirm the need for a new method. Concerning this new method, claims Hooke, no man has had any significant thoughts except the "incomparable Verulam." Bacon, however, did not say everything. Let us recall that both Bacon's *New Organon* and *New Atlantis* were left unfinished. Hooke proposes to continue Bacon's work with his own "Philosophical Algebra, or an Art of directing the Mind in the search after Philosophical Truths." [12] Hooke's proposed "algebra" was to be a set of simple rules on how to make fruitful discoveries about nature. But Hooke, too, left his paper unfinished.

In a similar passage, contained in some partially completed papers, appended by Waller to Hooke's "A General Scheme, . . . ," Hooke draws a parallel between mathematics and "Physicks" (i.e., experimental or natural philosophy). He states that, just as in geometry where conclusions are demonstrated in an orderly deductive manner based upon self-evident principles, so in "Physicks Geometrically handled" conclusions must be the result of an orderly process based upon principles "founded upon the immediate Objects of Sense disintangled from all the Fallacies of the Medium and Organ." [13]

For Hooke, "Physicks Geometrically handled" and Baconian induction (and, as he states elsewhere, the "Synthetick Method") are synonyms.[14] "Algebra," as Hooke himself tells us in the quote before the last, is an "Art

[11] P.W., p. 3.

[12] P.W., pp. 6-7.

[13] P.W., p. 73.

[14] Hooke's use of the terms Synthetick and Analytick was opposite that of Newton's. For Newton, analysis was making experiments and observations (i.e., induction), while synthesis was a deductive process in which explanations were drawn out of principles. For Hooke, the synthetic method was induction while the analytic was the deductive approach which Hooke regarded as secondary in importance to the inductive method. See Newton's *Opticks,* bk. III, query 31 and P.W., pp. 65, 173ff, 330-331. Also, in order to avoid confusion later, one should add that Descartes called the geometrical or deductive method synthetic, while the methodical search (of his *Meditations*) was analytic. See the end of his reply to the second set of Objections.

of directing the Mind." In the case of physics, however, Hooke uses the term algebra analogously. Hooke takes a mathematical term and gives it a new meaning. A mathematical algebra directs the mind in its search for mathematical truths. A philosophical algebra, on the other hand, directs the mind in its search for truths about nature. They are similar in that they both require a certain orderly procedure, an essential part of which was to have a foundation of indisputable truths. In geometry conclusions are deduced from self-evident *a priori* truths. In natural philosophy explanations are built up by induction from the self-evident *a posteriori* truths of sense experience. Without this wonderful inductive method of proceeding in natural philosophy there could be no consistently worthwhile results, even though one might occasionally hit upon a truth by accident. Therefore, in theory at least thought Hooke, one had to build up a vast structure of undeniable truths or "histories" about nature before any real progress could be made.

As stated above, Hooke preferred the "synthetick" method in his explorations of nature. But what of the "analytick" method? As is stated in his *Posthumous Works* (pp. 83-84) at the beginning of one of his lectures on light entitled "Sect. II. *Containing the Lectures of Light read about* Michaelmas, 1680," the natural philosopher must *first* collect the data and *later* try explaining the how and why of the facts. According to Hooke, "This is the true Method of coming to the Knowledge of all the Operations of Nature, and therefore whoever goes the other way to work, and begins *a priori* to this first of the Cause, and then to deduce the Effects from it, as a great Man has done, or at least would be thought so to have done; begins at the wrong end, and at length when he came to the ultimate and most visible Effects, he found himself, or at least most Men have found it for him, that he was much at a loss and unable to get out, and extricate himself." The "great Man" was none other than Descartes.

In another place in his *Posthumous Works* (pp. 173ff) we learn why Hooke thought this way. The Curator had no *a priori* prejudice against this method. However, as he saw it, from the practical point of view it is not very effective when it plays the sole or predominant role. According to the analytic method, most of our knowledge about nature is to be derived from a small number of universal principles posited at the outset. Along the way, the various deductions can be checked by experimentation. If the results should closely approximate the experimenter's expectations, the deductions in question, as well as those that came before, all the way up to the highest principle, can be considered as verified.

Although Hooke mentions no one in particular, he undoubtedly had

Descartes in mind when he described and criticized the analytic approach. The main problem with the analytic method is its lack of certitude. This is so because there is usually a break somewhere along the line in the chain of deductions. One can never be sure that an important step has not been inadvertently left out or that some other premise could not explain the phenomenon in question as easily as the original one. Why then struggle with the analytic method more than is necessary when there is a better way? This better way is the synthetic approach, i.e., basically Bacon's method.

There are two significant points to be noticed with respect to Hooke's words. For one thing, it must be remembered that the Curator is speaking *theoretically*. Hooke is describing the ideal situation, the way things should be done whenever possible. We will find that in actual practice, however, Hooke was eclectic, usually using the synthetic method but sometimes using the analytic method, rather than rigidly sticking to the facts of sense data in every case. The truth of what we say will become clear as individual cases are discussed later. Secondly, it should be noted that when Hooke speaks about his philosophical algebra he is not committing himself in any way to the extensive employment of mathematics in his pursuit of truths about nature. There is no actual use of mathematics necessarily involved; he is merely drawing an analogy.

## 2. 3 Hooke and Baconian Induction in Practice.

We will now probe more deeply the eclectic character of Hooke's thinking. This can be accomplished by discussing both the Baconian (i.e., synthetic) and non-Baconian (i.e., what one might call the Cartesian or analytic) characteristics of Hooke's thinking as revealed in his actual work. This section will concentrate on the former. In order to avoid needless repetition, the latter will be seen *passim* throughout the remainder of our study.

First let us briefly compare Hooke's approach to nature with that of Bacon's in order to establish the fact that the *aims* of both men were essentially the same.[15] It cannot be denied that Hooke was deeply imbued with the spirit of Bacon's approach to reality. This can be seen in numerous places throughout Hooke's writings and in the records of his lectures and reports before the Royal Society. In Hooke's first known treatise, *An At-*

[15] A complete analysis of induction would be impossible here. We will restrict ourselves to a study, drawn from all areas of Hooke's interests, of what Hooke understood by Baconian induction as shown by Hooke in his actual work.

*tempt for the Explication of the Phenomena* (of Capillary Action), which is reproduced in his *Micrographia*, the reader is warned to avoid the error, pointed out by the "thrice Noble and Learned Verulam," of concluding upon insufficient evidence. In the spirit of Baconian philosophy, one must have constant recourse to experiments backed up by sense knowledge and be ready and willing to reject old theories as new facts are gleaned. When discussing the nature of the phenomenon of gravity, for instance, Hooke insisted that his hypothesis was founded "upon the Phenomena of Nature, and not taken up at random, or by chance." [16]

Later, as he was discoursing upon possible improvements in the barometer, he recommended using a long tube so that every variation could be noticed. He observed that many of the operations of nature are out of the reach of our senses. Nevertheless, "there is no method of information so certain and infallible, as that of sense, if rightly and judiciously made use of" in the investigation of nature.[17]

Five years later, in the course of another paper, he digresses a moment to plead, in regard to the then present situation in natural philosophy, that the "harvest is great, but the labourers are few; and without hands and heads too, little can be expected; and to rely only upon time and chance, is, probably the most likely way to have all our hopes frustrated." [18]

We again witness the spirit of Bacon breathing in Hooke's thinking as he speaks, in the following year, a Latin phrase which he regarded as a maxim to be found in both the Prophet Daniel and Lord Bacon; namely, *Multi transibunt et augebitur Scientia.* A short space above in the same work, the ghost of Bacon can again be heard whispering in Hooke's ear. Hooke was worried because he detected an attitude among the young men of his time which said that there was nothing more to learn by applying one's senses to nature, and, furthermore, that no monetary gain was to be gained by employing with renewed vigor the telescope and microscope. Hooke affirms the opposite. Many things, he insists, have yet to be directly experienced if the young men would but look.[19]

One sees in Hooke a living, breathing example of one of Bacon's Fellows as described in his *New Atlantis*. These ideal scientists were constantly attempting to wrest something from nature. Sometimes this could be done by simple vision. Sometimes it would be necessary to go beyond simple vision. When such a situation arises the scientist must have recourse

[16] P.W., p. 178.
[17] Gunther, Vol. 7, 2/3/1686.
[18] Gunther, Vol. 7, 12/1691.
[19] See Gunther, Vol. 7, 2/1692.

to various experiments which will reveal what ordinary experience keeps from us. This is not always an easy task. It often calls for much patience and special equipment. But the ideal natural philosopher must be prepared to face such difficulties and see his work through to the end. Such a person was Hooke. Although he occasionally had brilliant insights, his usual day to day role was that of a plodding, often stumbling, experimenter constantly working on nature to yield up her secrets to man.

Besides a general agreement in spirit with the incomparable Bacon, Hooke can also be seen to agree with Bacon on more particular points. Such contact points can be seen in Hooke's views on the role of mathematics in the interpretation of nature and the importance of contrary instances, crucial experiments, simplicity, and practicality with respect to the prying loose of nature's secrets.

Since the proper way to proceed in natural philosophy was by means of carefully collecting data via one's senses, and, since mathematics proceeds without having to have recourse to sense knowledge, Bacon had given a minor role to mathematics; made it a mere appendix to the substantial sciences whose body was built up by careful observations. One notes the same attitude in Hooke's work in natural philosophy. To the Curator's mind, mathematical abstractions are of little use to the scientist who is interested in getting at nature as it really is in the concrete. Let us take a typical example of Hooke's thinking on the subject.

At a meeting of the Society on 4 April 1678 the members were discussing various methods for measuring sea depths. In the course of the discussion several members criticized a method previously suggested by Hooke on the grounds of Galileo's law for falling bodies. On this basis, they said, the time for the descent and ascent of the device could not always be in proportion to the sea depth as Hooke claimed it would be. Hooke countered by affirming that a terminal velocity would be reached after two fathoms. The critics urged further that Galileo's law of falling bodies would not allow for a terminal velocity in so short a space. Hooke's rebuttal was that Galileo's conclusion was based upon a mathematical theory rather than upon experimentations whereas his approach was to deal with the world by way of experimentation.

This does not mean that Hooke completely spurned the use of mathematics. But in Hooke's hands the use of mathematical correlations (which he never formulated in modern notations) was always simple and always an auxiliary tool when applied to nature. We do not find in Hooke's natural philosophy anything comparable to the consistent use of long, complicated and highly precise mathematical formulations found in Huygens,

Newton, and to a lesser extent in Galileo. Furthermore, there is no hint in Hooke's work that he believed the book of nature to be written in the language of mathematics. At best, those parts of Hooke's prose writing, which can be summarized today in some modern mathematical notation, were but summations of data gleaned from experimentations. They are usually after the facts; rarely before the facts. To Hooke's mind, in contradistinction to Hooke's appraisal of Galileo's thinking, an emphasis upon mathematics would never allow one to *discover* anything fundamental and actually true about nature.

We see here a difference in degree. Hooke was a thinker who did employ mathematics to some extent but who, in comparison to such thinkers as Galileo, Huygens and Newton, cannot be considered a person who emphasized mathematics or who attempted (although there are a few exceptions) to deduce knowledge about nature by the use of mathematical formulas. The Curator's *relatively* non-mathematical approach was deliberately cultivated because of the following reasons. For Hooke the ideal scientist must go directly to nature with all his five senses straining to pick up every bit of information available. This is certainly not how the mathematician operates. A person can be an ideal mathematician without ever performing one experiment. Given *a priori* postulates the mathematician can do great things. But how would such a procedure ever contact the real, physical world in which we live and breathe? Of course it cannot, answers Hooke. As applied to nature a mathematical formula merely correlates various superficial measurements. It cannot penetrate to the very heart of things and tell us how nature really is in itself. A mere formula can tell us nothing about the actual mechanisms underlying the phenomena of nature. Consequently, since Hooke firmly believed that the kind of knowledge he wanted could not be obtained by mathematics, he neglected this approach in his philosophy of nature.

It might be queried why Hooke could not have established his first principles by induction and then proceeded to apply mathematics to them, as Newton seems to have done. This would make him a mathematical physicist since his principles would be physical while his mode of proceeding would be mathematical. His methodology would then be inductive toward the physical first principles and deductive away from them. The simple fact is, however, that, although he could have done so, he did not do things in this way. This does not mean that he opposed all deduction in the sense that he opposed all reasoning and relied instead upon a rigid and strict empiricistic approach. As we will see in a later section, Hooke did allow for the limited use of the hypothetical-deductive method, for analogical reasoning, etc., when his inductive methods could not be applied.

With respect to contrary instances in Hooke's outlook on scientific endeavors, the importance of which was so emphasized by Bacon, one can bring forward many examples which disclose the Curator's mind on the subject. Three instances occurring between 1679 and 1682 will illustrate this point of agreement between Hooke and Bacon. During this period, the Royal Society was interested in the relationship between air and the maintenance of life in animals. At the very beginning of 1679, Hooke proposed various experiments to test this relationship. A week later, Dr. Croone, a member of a committee appointed by the Society to discuss past experiments and consider future ones, presented his view as to why small animals enclosed in an airtight container soon die. On Croone's view, such a death was the result of the animal's being overcome by its own breath and body steams. Judiciously, however, Hooke noted for the group that if such were really the case, then how would one account for the fact that animals enclosed in compressed air survive for a longer rather than a shorter period of time.[20]

At the end of the same year, Hooke was attempting to demonstrate his theory of how the atmosphere increases and decreases in weight as indicated by barometer readings. According to Hooke's proposed explanation, the ether was impregnated with various other bodies, such as steams and vapors from the earth, thereby increasing its specific gravity while maintaining a constant volume. This view was opposed to that of Flamsteed's, the Royal Astronomer. Flamsteed believed the levity and gravity of the air to be a function of its motion. But how could such be the cause, Hooke argued, for often the barometer is low while strong winds are blowing and high on a fine, calm day. With such contrary evidence, Flamsteed's theory was far from being a certitude.[21]

On another occasion, members of the Society became involved in a debate centering around the "flame of life." Some saw fit to argue that because phosphorus had a shiny appearance and could be extracted from blood and urine it constituted such a flame. Hooke dissented. Obviously, he commented, merely because phosphorus is shiny does not prove a thing. In addition, it can be extracted from several non-living substances. Such evidence must lay to rest once and for all the notion that phosphorus constitutes *the* "flame of life."[22]

[20] See Gunther, Vol. 7, 1/9/1679.
[21] See Gunther, Vol. 7, 12/4/1679.
[22] See Gunther, Vol. 7, 2/8/1682.

Consider now Hooke's stand on the place of "crucial experiments" within the edifice of natural philosophy. Hooke has publicly mentioned such experiments on several occasions. Bacon had set out in book II, section 36 of his *New Organon* a list of twenty-seven "prerogative instances" designed to further distill the "first vintage" obtained from the three tables. The fourteenth prerogative instance was the decisive, judicial, or crucial instance. Such an instance was an experiment that would conclusively decide which of two explanations for some phenomenon was the true one.

Hooke's belief in the efficaciousness of crucial instances is exhibited throughout his career. About 1670 the Curator experimented with his perpendicular telescope in order to determine whether or not there was a shift in the angle of parallax for a fixed star. This, he thought, was a crucial experiment, which would finally prove the Copernican hypothesis. The outcome, for Hooke, was positive.[23]

It was in the Autumn of his life, while in a state of declining health and activity, that Hooke delivered a series of papers on the nature and origin of amber and once more affirmed his belief in the worth of crucial experiments. In the course of his third discourse on the subject, the aging Curator reported that since the last meeting he had read Thomas Bartholine's *Acta Hafniensia* (Copenhagen, 1673) in which it seemed he agreed with Hooke's view that amber was nothing more or less than petrified tree gum. Bartholine, in fact, had listed eleven facts supporting the tree gum theory. To these Hooke quickly added six of his own. With no contrary instances and in possession of a set of facts roughly comparable to Bacon's three tables, Hooke felt safe in declaring with respect to his own supporting evidence that "These, I confess, to me seem to be *experimenta crucis*, as the Lord Verulam says." [24]

Bacon's note of simplicity can also be found vibrating in Hooke's approach to nature. If there was one thing Hooke was anxious to avoid it was a complicated experiment, device, or instrument that could not be easily understood or manipulated. Let us return to Hooke's attempt to explicate capillary action. It was Hooke's opinion that the unequal heights of water columns to be seen in tubes of varying diameter were caused by unequal air pressures. As the tubes decreased in diameter, claimed Hooke, the air pressure inside became less and less while the outside pressure remained relatively constant thus forcing the liquid higher up the tube. This was

[23] See Hooke's *An Attempt to Prove the Motion of the Earth*, first published in 1674 and reprinted in Gunther, Vol. 8, pp. 1-28.

[24] Gunther, Vol. 7, 5/19/1697.

caused by the greater affinity of water for glass than of air for glass. To prove this, Hooke decided that he had to prove the truth of two propositions. "The first of which is, That an unequal pressure of the incumbent Air, will cause an unequal height in the water's Surfaces. And the second is, That in this experiment there is such an unequal pressure." [25]

The first proposition was easily demonstrated by blowing and sucking on the open ends of a U-shaped tube partially filled with water.

To prove the second proposition he designed a long glass tube fitted with a small bowl and various sized tubes, to be described later. On the basis of this simple experiment, which anyone could perform, he thought that he had proven his equally simple hypothesis. Unequal air pressures, he claimed, "is a cause sufficient to produce this effect, without the help of any other concurrent; and therefore is probably the principal (if not the only) cause of these Phenomena." [26]

Later in the *Micrographia,* when discussing the nature of the air in relation to other phenomena of nature, Hooke reaches the conclusion that many of the most mysterious phenomena of nature, such as the changing shape and size of the setting sun, can be explained by knowing about the air. What need is there for long and complicated explanations when only two simple propositions are sufficient, asks Hooke. Given the medium of the air, and the fact that the density of the air will vary from place to place, he can explain everything.[27]

Later in life, Hooke showed to the Royal Society two ways in which a horizontal circular motion could be converted into an angled motion of anywhere from 0 to 90 degrees by the use of notched gears. Again he emphasizes that simplicity is the key note of his inventions. He states that, "contrary to the opinion and practice of most projecting mechanics and ignorant spectators," the simpler a machine is the better it is.[28]

About the same time, Hooke showed a new scale of his own invention which could determine the decimal, centiesimal, or millessimal fractions of any given weight. He called it his "proportional balance" and pointed out its extreme simplicity. Yet, no one had thought of it before. This was due to its being "altogether as obvious, as to set an egg on end." [29]

In 1691, when reading a paper on a method for sounding the depths of the sea, he described a method for making a device which would float on the surface and record the distance of the sounding device as it came up.

[25] M., p. 11.
[26] M., p. 21.
[27] See M., pp. 217ff.
[28] Gunther, Vol. 7, 7/18/1683.
[29] Gunther, Vol. 7, 12/5/1683.

He interrupted his paper to explain to his audience that he realized that things were becoming a little complicated but that such was certainly not his intention. He promised to continue his efforts at keeping everything as simple as possible so that "any one, that can but write and read, can be able to make trial therewith, and keep account thereof." [30]

Finally, that Hooke's mind, like Bacon's, was constantly turned toward the useful and practical is a fact which so pervaded his work as to hardly be in need of a detailed substantiation. At the outset of a discourse delivered in 1695 on measuring heights and distances at sea, Hooke once more reiterates and emphasizes how the present discourse, like everything else he saw fit to undertake during his long and active career, will ultimately prove most useful to mankind. This is as it should be for man's rightful place is as the master of nature and his environment.[31]

As we have seen, Hooke and Bacon are kindred spirits. There are no basic differences between Hooke and Bacon on the question of the ideal methodology. In addition, it might be noted, the two men have enunciated similar stands with respect to scientific and cosmological conclusions in at least two instances. Following his discussion of the three tables, Bacon defined heat (which is one of the only two concrete results of his new organ that he lists)[32] as an expansive motion, tending to ascend, extending through diminutive parts of bodies, rapidly checked, and driven back and forth as seen in the two pre-eminent examples of flames and boiling water. Similarly, Hooke tells us in his work on capillary action that heat is nothing but a brisk and vehement agitation of the parts of bodies. Also, neither the Curator nor the Viscount thought very highly of the atomists. Hooke complained about the "anatomists" who said atoms have definite and stable figures, called a vacuity or empty space an "imaginary Entity," and criticized the "Epicurean atoms" as being incapable of explaining weightiness.[33] For his part, Bacon, when discussing his methods which should be employed for investigating latent properties of bodies, declared that "This method will not bring us to atoms, which takes for granted the vacuum, and immutability of matter (neither of which hypotheses is correct), but to the real particles such as we discover them to be." [34] One wonders if Bacon could have read Scotus for whom a concrete thing was a concrescence of simpler "formalities."

[30] Gunther, Vol. 7, 12/16/1691.

[31] See Gunther, Vol. 7, 2/13/1695.

[32] The other is whiteness: a mixture of two transparent bodies with a certain simple and uniform disposition of their optical parts. See A.L., III, 4.

[33] See P.W., pp. 172ff, 191.

[34] N.O., II, 8.

*2. 4 Hooke's Development of Baconian Induction.*

We have observed how Hooke looked upon Bacon's scientific methodology as an ideal. It must now be pointed out that Hooke could not always live up to this ideal. He did in fact add elements of his own and at times deviate from the basic Baconian outlook.

Among his contemporaries, Robert Hooke was known as an assiduous and conscientious experimenter. Even Newton, with whom he often quarreled, credited Hooke with being a "curious and accurate experimenter" in a letter to Collins dated 10 December 1672. Hooke, however, did not operate under any false illusions. Experimental verification was of great importance but it was not everything. In fact, in his "To the Reader" at the beginning of his *Lectiones Cutlerianae* he makes it clear that a lifetime could be devoted to experimenting on any one subject but even then the subject would not be exhausted. Also, he does not hesitate to admit that simple luck and chance is often responsible for a successful result. As in multiplication, trial and error is important.

But how did he actually carry out his experimentation? This involved various factors: the recognition of a need for universal standards and international cooperation among scientists, the use of controlled experiments, the use of hypotheses that were based on facts and necessary at the time of use, the use of analogical reasoning, and knowing when to explain away experimental errors.

The need for universal standards was considered by Hooke to be an important aspect of experimentation. Such standards would facilitate not only the work of individuals but also the mutual communication and understanding among individuals and groups. This was one of the reasons for his interest in the motion of pendulums. He tells Boyle in a letter relating the activities of the Royal Society how the "experiments we are now most busy about, are concerning the adjustening of the length of pendulums, thereby to settle a common standard for length." [35] It was almost exactly at the same time that Hooke read a paper on universal measure before the Society. The paper gives Hooke's views on the general criteria for the universal measures of length and time. The universal measure of length, it is stated, should be of one, pure, homogeneous, metallic substance in some kind of prismatical shape. This would not be so hard to accomplish. In the case of time, however, one runs into all sorts of difficulties. Hooke suggests a short pendulum but quickly lists the shortcomings of such a device. The two foremost difficulties are the lack of constant

[35] Letter from Hooke to Boyle 13 December 1664.

motion of any known pendulum and the fact that the earth's attraction may vary from place to place thus altering the pendulum's swing.[36]

As time went on, Hooke became more precise in his suggestions. In 1670 he presented two ideas for a universal measure of length. One was to drop mercury on a metal plate and consider the spread of a determined number of drops to be the standard inch. The other was to do likewise using distilled water instead of mercury.[37] Still later he expressed the view that a drop of mercury would make a good universal measure of weight.[38] None of the above suggestions was ever implemented. Nonetheless, Hooke never lost sight of the importance of universal standards. In the middle of 1683, to mention one statement on record, Hooke outlined an experiment to show how the true and comparative expansion of any metal may be found. An iron weight, suspended from one pan of a balance, is forced under melted lead by adding weights to that pan. The iron would then be removed and submerged under other things and weights added to the other pan until they balanced. Then, assuming there was nothing more dense than melted lead, one would now know the comparative specific gravities of various substances. Although it is doubtful that Hooke ever actually completed the project, it is significant that such projects were stressed as a means for obtaining universal standards, quantitatively speaking, without which little progress could be made in natural philosophy.[39]

The importance of what today are known as controlled experiments was also recognized by Hooke. Very early in his career, Hooke showed an aptitude for employing controls upon his experimental undertakings. When, toward the end of 1662, he was endeavoring to determine the quantitative rarefaction of air under varying pressures, he not only used water but also "coarse spirit of wine" in this apparatus.[40]

Six years later, Hooke was experimenting on bodies falling *in vacuo* and suggesting experiments to test the springiness of bodies.[41] First he constructed a long glass tube out of which he pumped the air. He admitted, it might be noted in passing, that he thought the vacuum not to be too good. He then released a feather in the exhausted tube and found it required four seconds to reach the bottom. The experiment was then repeated with an unexhausted tube. The feather was found to require six seconds for its descent in the latter case.

[36] See Gunther, Vol. 6, 12/14/1664 and P.W., p. 472.
[37] See Gunther, Vol. 6, 1/20/1670.
[38] See Gunther, Vol. 7, 2/10/1686.
[39] See Gunther, Vol. 7, 7/4/1683.
[40] See Gunther, Vol. 6, 12/10/1662.
[41] See Gunther, Vol. 6, 10/29/1668.

Newton is credited with having emphasized the *hypotheses non fingo* dictum. The same can be said of Hooke. Robert Hooke was as adverse to conjuring up explanatory principles as he was to complicated and ostentatious experiments and devices. At the beginning of his *Micrographia,* as already mentioned above, Hooke praises the methodology of Bacon and holds it up as the banner around which all true seekers after truth in nature should rally. After giving his conclusion as to the cause of capillary action, Hooke ventures into various possible extensions and applications of his discovery to other phenomena of nature. In the course of his eighth query he digresses a moment to assure his readers that he is not engaging in merely idle speculation. "For I neither conclude from one single Experiment," states Hooke, "nor are the Experiments I make use of, all made upon one subject: Nor wrest I any Experiment to make it *quadrare* with any preconceiv'd Notion . . . so will all those Notions be found to be false and deceitful, that will not undergo all the Trials and Tests made of them by Experiments." [42]

[42] M., p. 28. Those espousing the new experimental philosophy in the seventeenth century understood the word hypothesis in a manner different from that generally understood today. For them *a priori* hypotheses were unacceptable, while *a posteriori* hypotheses were acceptable. They tended to think of all hypotheses as *a priori,* while we tend to think of them as all *a posteriori*. What they were opposed to was a typical Aristotelian way of arguing which may have had and still may have value in dealing with the ultimate questions of philosophy but which was not so fruitful in what today we call the natural or physical sciences. This method was to list all the possible alternatives or hypotheses supposedly solving a certain problem and then, by some process of ratiocination, eliminate the false ones until the truth was arrived at. The Baconians, on the other hand, wanted to collect data and arrive at the truth directly. Thus, it became acceptable to propose a hypothesis after the facts but not before. Newton neatly summarizes this attitude for us at the beginning of a brief letter sent to Oldenburg on 8 July 1672 as part of a series of letters he wrote attempting to defend his theory of light and colors. "In the mean while give me leave, Sir, to insinuate, that I cannot think it effectual for determining truth, to examin the several waies by which Phaenomena may be explained, unless where there can be a perfect enumeration of all those waies. You know, the proper Method for *inquiring* after the properties of things is, to deduce them from Experiments. And I told you, that the Theory, which I propounded, was evinced to me, not by inferring 'tis thus because not otherwise, that is, not by deducing it only from a confutation of contrary suppositions, but by deriving it from Experiments concluding positively and directly. The way therefore to examin it is, by considering, whether the Experiments which I propound do prove those parts of the Theory, to which they are applyed; or by prosecuting other Experiments which the Theory may suggest for its examination. And this I would have done in a due Method; the Laws of *Refraction* being throughly inquired into and determined before the nature of *Colours* be taken into consideration. It may not be amiss to proceed according to the *Series* of these *Queries*; which I could wish were determined by the Event of proper Experiments; declared by those that may have the curiosity to examin them." (*Philosophical Transactions,* Vol. 7 (1672), p. 5005 [misnumbered as 4004].) Even *a posteriori* hypotheses, though, can be

At the end of the *Micrographia,* under the title of "Observ. LX. Of the Moon," Hooke added, in order not to let an empty space in Plate 38 go to waste, some observations on how the surface of the moon was formed. These observations were made in October of 1664 with a thirty foot telescope. These opinions of Hooke's are interesting because, although he could not test them, they document for us the approach to nature he was attempting to maintain. Hooke claimed that the moon appeared to have short, shrubby vegetation growing over its surface and (N.A.S.A. take note) that its many pits were the result of internal pressures pushing up through the surface like earthquakes and volcanos on earth. The Curator believed this to be a reasonable explanation in lieu of any evidence that the surface was, in the past or presently, being bombarded by missiles from space. Also, he did not think the surface was soft enough to admit of such an explanation. Hooke was quite willing to admit such possibilities just as Ptolemy was willing to consider the possibility that the earth rather than the heavens moved. In the end, however, Hooke refused to allow a bombardment theory "for it would be difficult to imagine whence those bodies should come; and next, how the substance of the Moon should be so soft." [43]

A disrespect for all those whose theories were either not founded on facts or which contradicted one or more facts is most clearly seen in the works Hooke produced when at the height of his career in 1682. In one place, taking a highly critical attitude toward his contemporaries, Hooke maintains that comets, in order to explain their light, speed and retrograde motion, must be somewhat starlike. What other explanation is there that fits the facts? Some, notes Hooke, even despair of finding answers based upon the natural course of events and end by bringing in a *deus ex machina.* Others offer to explain a comet by means of fanciful causes that fail to account for the data. "Those that hold solid orbs," remarks Hooke, "will afford it no room, nor those that hold Vortices. Those indeed that suppose Dimons," the Curator adds snidely, "may suppose what they will, but to little purpose." It would appear that it was Copernicus' mental trait of not inventing wild and unnecessary hypotheses that endeared Copernicus and his system to Hooke. All other theories, believed Hooke, are full of too many contrivances and inconsistencies.[44]

With respect to the type of efficient causality responsible for the motions

later altered or replaced with the addition of more facts. See Newton's fourth rule for reasoning in philosophy (i.e., the natural sciences) at the beginning of bk. 3 of his *Principia.*

[43] M., p. 243.

[44] P.W., p. 167.

of the planets, Hooke also encountered various views which greatly displeased him. Without naming any persons in particular, except Kepler and Descartes, the Curator briefly lists some of these opinions for his readers. Some, he reports, postulate spirits or Intelligences or other such extravagancies to push the heavenly bodies around. Others maintain the existence of solid, crystalline spheres or epicycles moving around and carrying the planets with them or some other kind of "wheel-work." Kepler, and those that follow him, emphasizes the force of magnetism and also postulates the existence of friendly and enemy sides to bodies which involve them in some kind of anthropomorphic warfare. In addition, they talk about a real moon hidden within an outer, visible shell while also inventing radiating spokes of light from the sun which are supposed to help sweep the planets onward. But all this is utter nonsense cries Hooke. Descartes, he continues, has his whirling ether around each body. But these also are silly.[45]

Hooke also prosecuted his program of eliminating feigned hypotheses in areas other than those of astronomy and atmospheric pressure. During his second discourse on amber, delivered six years before his death, Hooke admitted that most authors on the subject disagreed with his tree resin theory but that this did not discourage him. He found strength in knowing that his view was based upon facts whereas the views of others were not. Hooke regarded Philippus Jacolus Hartmann, who had written a popular work entitled *Succini Prussici Historia Physica et Civilis,* as one of his main adversaries. Hartmann claimed that amber originated in the seas and was found on land only because it had been carried there by a great number of mysterious underground channels which crisscrossed the entire terrestrial globe. Hooke, on the contrary, thought amber to be the gums of trees washed down to the sea after petrifying. The reason why amber, like other things, is found in sand beds is because such are the remains of the sea which once covered the land. "I did," reported Hooke, "thirty-three years ago, prove, by multitudes of observations (divers made by myself, and many more by others) that all England is a most evident instance and testimony of the like phenomena here." In the future, continued Hooke, "if God restore my health, I hope I shall be able to give a more particular, convincing and satisfactory account." In any event, argued the Curator, his position should be preferred to Hartmann's if for no other reason than that it was simpler and did not conjure up all sorts of strange subterraneous conveyances in order to account for the origin of amber.[46]

Granted, therefore, that Robert Hooke would not accept what he con-

[45] See P.W., pp. 178-179.
[46] See Gunther, Vol. 7, 2/24/1697.

sidered to be unfounded assertions, what criteria did he employ in separating the well-founded from the ill-founded hypothesis? A proof of some theory in natural philosophy could take one of two forms for Hooke. One method is by direct sense knowledge. The other is indirectly by analogy. Hooke's analogical method can also be called his hypothetical-deductive method. That is, he did not hesitate to set up hypotheses to be tested when straight inductions were impossible. Such hypotheses, however, must always be based directly on experience, as we have seen.[47]

The direct method of proof is the best but unfortunately it is often unobtainable. An example of a case in which Hooke believed he had achieved such a direct demonstration is found in his *Micrographia.* Hooke is attempting, in one place, to defend the thesis that distortions in the appearances of distant objects were due primarily to light rays from those objects being bent by the intervening medium. To establish his position he had to prove two things. One, that light rays are indeed bent by the media through which they pass. And secondly, that our atmosphere possesses the properties of light-bending media. The first step is established directly; the second by analogy.

To establish the first step, Hooke constructed a rectangular box with the long sides made of glass. The bottom half of the container was then filled with a strong saline solution. The top half was filled with fresh water. Light rays were then allowed to enter the waters by passing over one end of the container and the shape of the shadow was noted. As expected, the hypotenuse of the right angled shadow was not straight but curved downward to a noticeable degree. "But that," Hooke presumed, "I have by this Example given proof sufficient (viz. ocular demonstration) to evince, that there is such a modulation, or bending of the rayes of light, as I have call'd inflection, differing both from reflection, and refraction (since they are both made in the superficies, this only in the middle)." [48]

The second part of Hooke's problem is to extend the results of his experiment to the atmosphere. This he does by heating the air in a glass bubble, about six inches across, by heating and sealing the glass bubble and then viewing an object through the rarefied air in the bubble. An object seen in such a way appears differently than it would if viewed through regular air. This proves to Hooke that changing the density of the air af-

[47] M. B. Hesse also recognizes the fact that Hooke was not opposed to postulating hypotheses to be tested. This may, in fact, have been the core of what he would have proposed if he had written something on his philosophical algebra. See her "Hooke's Philosophical Algebra," *Isis,* Vol. 57 (1966), pp. 67-83. The article, however, suffers by failing to emphasize Hooke's strictly Baconian side.

[48] M., p. 221.

fects the inflection of light rays. Now, by analogy, and with the support of barometer readings taken at different altitudes, which indicate varying air densities, he can affirm that what happens in the restricted context of his experiments also happens at large in the atmosphere.

Other examples of Hooke's trust in analogous reasoning, with respect to natural philosophy, are many and varied. On 28 March 1666, for instance, Hooke reported to the Royal Society concerning his observations on the moving spots on the surface of Mars. He also noted such spots on Jupiter. By analogy he concluded that Mars and Jupiter rotate on their axes just as does the earth.[49]

Later in life, while presenting his own thesis on the nature of comets and gravity, Hooke suggested that the flaming tails of comets might be composed of some substance between a solid and a fluid. However, he doubts that such a kind of body will ever be satisfactorily demonstrated since there is no way of obtaining any direct sense knowledge of it as the comets speed about through the ether and, furthermore, there is nothing analogous to such a body with which we are familiar and which could be likened to the comets.[50]

This was not the case with everything concerning the nature of comets, however. In the same work, in order to illustrate the kind of thing he had in mind, Hooke claimed to have experimented with a little combustible ball suspended from a wire. After lighting the ball and swinging it through the air, Hooke reported that the appearances were so close to those observed in the heavens that he felt safe in affirming, by analogy, that indeed his little model was a comet in miniature. The effects directly observed were due to a hard center burning away with the flames and smoke being borne aloft by the air. By analogy, a comet is a hard core set afire with effluvia being borne away by the ether. If someone should object, interjects Hooke, that his experiment was all well and good but *omne simile non est idem,* Hooke would answer that such is true but, nevertheless, it is a much better explanation, based upon facts, than anyone else had put forward. As far as he was concerned, it had been demonstrated. Perhaps not with absolute certitude, but at least sufficient for the purposes of natural philosophy.[51]

A bit later, the Custodian of Experiments stated that sensible effects are known through experiments and observations. When dealing with insensible effects, however, one must have recourse to another method

[49] See the *Philosophical Transactions,* Vol. I (1665-1666), pp. 3 and 239-242.
[50] See P.W., p. 166.
[51] See P.W., p. 167.

which makes them known through the "Probability from Similitude, Harmony and Uniformity in the Operations of Nature." [52] Also, Hooke insists upon the roundness of the earth so he can, he says, argue by analogy to all globular bodies when he finally states his gravitational theory.[53] Two pages further on, Hooke comments that, since he had already shown that the earth moves around the sun and that we, therefore, are one of the celestial bodies, he can safely extend by analogy whatever he discovers about the earth to the other celestial bodies.[54] We will see other examples of Hooke's analogical approach in succeeding chapters.

One may well wonder, at this point, to what extent Hooke remained rigidly attached to his standards and how strictly he applied his methodological pronouncements. The answer is, he sincerely tried to operate in accordance with his principles but, nevertheless, on various occasions he found it necessary to mitigate his stand and allow for a certain margin of experimental error. At the end of 1662, for instance, after performing an experiment designed to show that the degree of rarefaction of air and the force exerted by that air are in a reciprocal proportion, Hooke admitted that his results did not bear out his hypothesis. He did not give up the theory, however. Rather, the discrepancies between what he expected and what he obtained were explained away on three counts. First of all, one of the tubes used in the experiment was not of uniform diameter throughout. Secondly, there was an unknown amount of impurities in the air used in the experiment. And finally, previous trials had come out much closer to the anticipated results.[55]

Later on, in the Curator's lecture on proving the annual motion of the earth, one comes to what seems to be an unique case with respect to Hooke's attitude toward the non-confirmatory results of some of his experiments. In 1670 we find him saying *beforehand* what he will say if his experiment did not show the anticipated results. He remarks, at the beginning, that many learned people do not admit the Copernican theory. Some, such as the Jesuit astronomer Riccioli, even go so far as to give all sorts of arguments against the heliocentric hypothesis. Nonetheless, claims Hooke, his observations will answer all and every objection both old and new. And, even if he should fail to find some angle of parallax for some star, he would maintain his belief in Copernicus' essential correctness by affirming the almost infinite extension of the universe. This dis-

[52] P.W., p. 172.
[53] See P.W., p. 178.
[54] See P.W., p. 180.
[55] See M., pp. 222ff.

closure is all the more remarkable when we note that in the same part of the same work he called his parallax observations an *experimentum crucis*. It would appear that Hooke was sometimes willing to place a beautiful hypothesis or theory above the results of any one experiment.

If one has any doubts about his willingness to do so, they quickly evaporate when one reads Hooke's treatise on comets and gravity. In the course of his 1682 disquisition Hooke lists nine properties of gravity which any attentive person can observe. According to the ninth, heavy bodies should decrease in weight as they are elevated above the earth. Hooke inserts here that he had tried to prove this by various experiments in Westminster Abbey, St. Paul's Cathedral, Banstead Downs, etc., but without success. Nevertheless, he tells his readers that he is inclined to believe that his experiments were faulty since a decrease in weight is a necessary part of his theory of gravitation.[56]

It is important to note these discrepancies with respect to Hooke's own basic dictum of sticking to the facts so as not to be shocked by some of his other pronouncements. Rather than being an absolutely strict inductivist he sometimes did employ untested theories. As we will shortly observe, Hooke, in addition to those cases mentioned above, did not hesitate to accept such important doctrines as the "mixture theory" of air, the ether and Descartes' three laws of motion, on a non-inductive basis. But why did Hooke deviate from his beloved synthetic method on occasion? Certainly he was influenced by his teachers to accept doctrines dear to them. Certainly the aesthetic attraction of certain beautiful, all encompassing theories had its effect upon him. But of most importance, as we will see, is that Hooke felt that he needed certain theories to explain hosts of facts which he felt could not be explained on any other basis. At certain times, then, Hooke tended to be eclectic, using the Cartesian, deductive or analytic approach rather than a strictly synthetic approach.

In fine, we find Hooke developing Bacon's basic inductive method by adding some precision to the experimental procedure. He realized the need for exact measurement and international cooperation in science. He attempted to state simple conclusions and explanations as well as keeping to an over-all methodology that was simple to use. However, he did not spurn the use of hypotheses and analogies where direct observations were impossible as long as they did not contradict the facts. By these methods, one could at least eliminate wrong conclusions even if one could not have absolute certitude with respect to the truth of his hypotheses. But, after all, Hooke was only human and so, despite his several self-imposed safe-

56 See P.W., p. 182.

guards, could not resist on rare occasions making the facts fit the theory rather than *vice versa.*

At this transitional point between the background material pertinent to Hooke's work in mechanics and our consideration of his actual attempts, it would be well to know what Hooke himself would have understood by mechanics. As we learn from the *New English Dictionary* (subtitled *On Historical Principles*) there was in the seventeenth century a multitude of meanings for mechanics. Bishop Wilkins, a patron of science and a friend to Hooke, stated in his *Mathematical Magick* (London, 1634) that "Astronomy handles the quantity of heavenly motions, Musick of sound, and Mechanics of weights and powers." According to Aubrey, Wilkins made Hooke a gift of this work, which Hooke greatly appreciated. Robert Boyle at the beginning of his *Of the Usefulness of Mechanical Disciplines to Natural Philosophy* (London, 1663), distinguishes between two meanings. One, the more proper, calls mechanics that doctrine which touches upon those forces involved in moving bodies and also the making of engines which would multiply human power when it came to doing work. The other, the one Boyle said he would use himself, calls mechanics all "those disciplines that consist of the applications of pure mathematicks to produce or modify motion in inferior bodyes." Hooke never clearly stated his definition of mechanics. However, in most cases where it appears he seems to be using it in Boyle's more proper sense. Hooke, living before the age of specialization, can be forgiven for his ambivalence. For Hooke, mechanics was more of an approach than a separate subject heading. What he was approaching was natural philosophy, a grand collection of arts and sciences including just about everything but theology and the professional fields.

## CHAPTER III

# THE MECHANICS OF FLUID MOTIONS

### *3.1 The Legacy of Bacon and Boyle.*

For Hooke, there were many problematic areas concerning air. The nature of the air itself, however, as we will see, was not a problem. The problematic areas, of which there were as many as one could find things affected by the air, were the effects of air on other things. Because they greatly affected human existence, these issues were important to Bacon, Boyle and Hooke.

In his *New Organon* Bacon calls the twenty-first rank of the Prerogative Instances the "Instances of Completion."[1] He notes here that actions take place in limited and definite spaces. When investigating any phenomenon, thought Bacon, it was very important to also investigate the medium or space in which the phenomenon transpired because the medium might have some effect upon the phenomenon. The most obvious example is the occurrence of various phenomena in the all-pervasive medium of the air. Therefore, an investigation of the air becomes important.

Later, in his twenty-sixth rank of Prerogative Instances called the "generally Useful Instances," Bacon again brings up the topic of air. Because "common air" is always around us we must be prepared to examine various ways in which it may help or hinder us. There are two general ways of doing this. One is to figure out ways of excluding the air from something while the other is to devise means of keeping the air in. With respect to the former, one must investigate different ways of making airtight containers. One might also keep air out by surrounding something with powder (although powder is not so useful since it contains air) or by putting it under water (apparently water, unlike powder, does not contain air). On the other hand, when it comes to keeping air around some-

[1] See N.O., II, 45.

thing, one should experiment with an inverted tub forced under water and hermetically sealed containers.

In the last two paragraphs we have seen Bacon say two things. First, because the air is so all-pervasive it must be given much experimental attention. Secondly, in order to study the air it must first be isolated. Why Bacon wanted to study the air seems clear. He wanted to know whether it was relevant to the causal actions of bodies immersed in it. There is a difference between asserting that p alone is the cause of q and asserting that p and air, together, are co-causes of q. We also find this same attitude in Boyle and Hooke for whom there is no one problem of the air. Rather, anything involving air was fair game for pneumatical experiments.

Bacon's methodology and spirit of curiosity had a direct influence on Robert Boyle.[2] As it has been observed, from the time he was twenty (1647), Boyle came under the influence of a group of men meeting in London to study the New or Experimental Philosophy. In 1654, Boyle moved to Oxford and became a part of the John Wilkins circle. All its members were passionate believers in Bacon's methodology. As we know from his own work, Boyle also became an ardent follower of Bacon. And, as a part of the Baconian legacy that rubbed off onto Boyle, one finds an interest in the air.[3]

Soon after Boyle's arrival in Oxford he met Hooke. In his laboratory near University College, Boyle was busily at work attempting to design and construct some kind of device that would allow him to expand and compress air at will.[4] This would allow him the control over the air he needed in order to carry out "pneumatic" experiments. By being able to increase or decrease the amount of isolated air he could note what would happen to things in the enclosed, isolated area. Boyle set his new assistant, Hooke, to work on the problem and, about 1658, a reasonably efficient "air-pump" was produced.

The use of Boyle's new "pneumatic engine," however, did not substan-

[2] For a general study of Boyle's life and work see Louis T. More, *The Life and Works of the Honourable Robert Boyle* (Oxford, 1944).

[3] For a statement of the social, cultural, and religious background to Boyle's work, as well as the influence of Bacon, Torricelli, and Pascal, see H. Butterfield, *The Origins of Modern Science* (New York, 1962), ch. 7 and J. B. Conant, *On Understanding Science: An Historical Approach* (New Haven, Conn., 1947), ch. 2. For a general survey of Boyle's work on the air refer to N. Mohler, "The Spring and Weight of the Air," *The American Physics Teacher*, Vol. 7 (1939), pp. 380-389. For a more detailed account of Boyle's work in pneumatics see J. B. Conant, "Robert Boyle's Experiments in Pneumatics," *Harvard Case Histories in Experimental Science* (ed. by J. B. Conant and L. K. Nash, Cambridge, Mass., 1957), Vol. I, pp. 3-63.

[4] See M. Boas, *Robert Boyle and Seventeenth-Century Chemistry* (Cambridge, England, 1958) for the details of Boyle's work.

DIAGRAM ILLUSTRATING HOOKE'S AIR PUMP

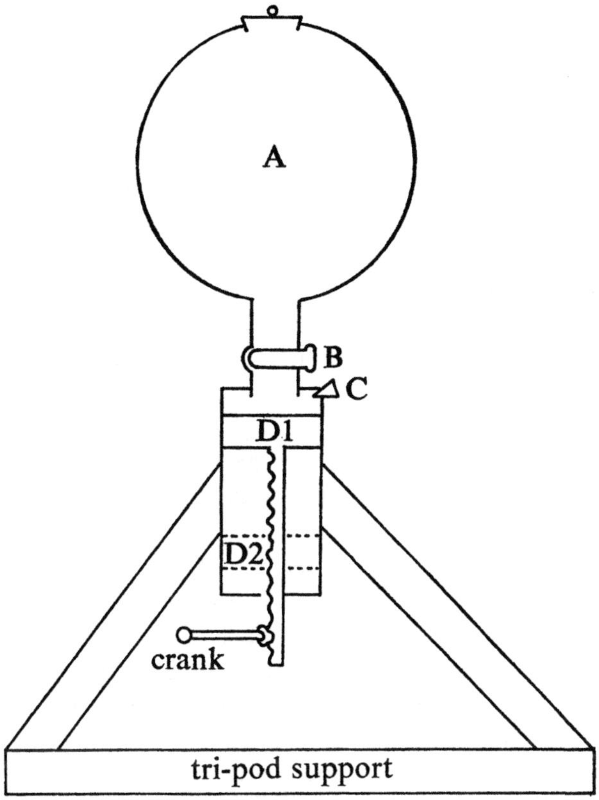


For rarefaction of air in A:

1) Open valve B; move D1 to D2; close valve B.
2) Remove valve C; move D2 to D1; close valve C.
3) Repeat steps 1 and 2.

For condensation of air in A: Reverse process.

tially alter anyone's conception of air during the seventeenth century. Everyone of note during the later seventeenth century went along with Clave's 1641 work entitled *Nouvelle Lumiere Philosophique* which held for a "mixture theory" of air, i.e., that air was a collection of vapors suspended in the ether.[5] This remained the basic attitude of both Boyle and Hooke throughout their lives even though, in retrospect, we can observe

[5] See *ibid.*, ch. 6.

in some of their experiments the beginnings of the modern view on the role of the air in respiration and combustion.[6]

This can be seen in Boyle's definition of air. Air, he believed, was a "confused aggregate of effluviums from such different bodies, that, though they all agree in constituting, by their minuteness and various motions, one great mass of fluid matter, yet perhaps there is scarce a more heterogeneous body in the world."[7] The numerous bodies making up the bulk of the atmosphere come from the earth.[8] The gases, fumes, vapors, etc., made "fluid" by the looseness of their parts, float in the ether and are literally pushed around by the unequal weights of different vapors suspended in different sections of the ether.

Also, because of its heterogeneous nature, the air was especially susceptible to movement. Accumulations of heavier vapors would push into lighter ones only to be dispersed by collisions with other types of particles. The heat of the sun could expand and lighten vapors; cold would contract them. The slightest change in any one part would affect all parts so that the degrees of condensation and rarefaction of the air over any one place were constantly changing.

Hooke shared Bacon's view on the importance of air and Boyle's view on its nature. He could not understand how anyone could question what he had learned from them. His attitude is epitomized by the following example. On one occasion some members of the Society expressed criticism of what they considered to be an undue emphasis upon air. Hooke was embittered by such remarks. Did these critics think that he could learn about the effects of air in an *a priori* way or by revelation? He affirmed that the exhausting and condensing of air was no trivial trick. On the contrary, "an exact and thorough knowledge of that is of more concern to

[6] According to Boyle, air is fit for respiration because it contains a certain "quintessence" or "spirituous part" which, when pumped out, leaves behind only the heavier, grosser parts that are unable to "cherish the vital flame residing in the heart." See Boyle, Vol. I, p. 69.

In a similar fashion, Hooke believed that respiration, as well as combustion, depended upon certain bodies in the air which, when removed or somehow "satiated," for example, when a small animal was kept inside an airtight container long enough, brought on suffocation. See Gunther, Vol. 7, 1/9/1679.

For a discussion of the problems of respiration and combustion with respect to the nature of the air see T. S. Patterson, "John Mayow in Contemporary Setting," *Isis*, Vol. 15 (1931), pp. 47-96; 504-546 and D. McKie, "Fire and the Flamma Vitalis: Boyle, Hooke and Mayow," *Science, Medicine and History* (ed. by E. A. Underwood, Oxford, 1953), Vol. I, pp. 469-488.

[7] Boyle, Vol. 3, p. 463.

[8] See Boyle, Vol. 5, p. 111. See also Vol. 4, pp. 25, 27, 31.

mankind than all the other physical knowledge in the world." Everything that exists on the surface of the earth and above the earth depends in one way or another on air. It is the *sine qua non* of life itself. "Infinite and unspeakable," Hooke concluded, "are the uses of it to the husbandman, the merchant, the tradesman, the mechanic, etc. And that age will be deservedly famous, which shall perfect the theory of it." [9]

*3.2 Capillary Action (1661).*

About 1660 Boyle observed something which baffled him but which he did not attempt to explain. This was the fact that, other things being equal, the finer the tube, the higher up in it a fluid would rise. This phenomenon is today called capillary action. A year later Hooke published a small treatise entitled *An Attempt for the Explication of the Phenomena Observable in an Experiment Published by the Honourable Robert Boyle.* This work was later made a part of Hooke's *Micrographia.* Hooke's separately published treatise (1661) and its reproduction in the *Micrographia* (1665) mark the beginning and end of Hooke's published work on this topic. There is no development in his view, the latter work being simply a restatement of the former. In our examination of his explanation of the phenomenon observed by Boyle we will have reference to the latter work.

The Curator's problem was twofold. On the one hand, he was interested in explaining only the case of liquids rising in fine tubes. On the other hand, he was interested in the broader question of the relationship between liquids and things with which they may come in contact. Hooke felt that if he could successfully explain the specialized case of the relationship between the water and the fine glass tubes he might also be able to handle such seemingly diverse cases as the absorption of liquids by lamp wicks, sponges, blotters, and plant roots; the rising of sap in trees; the rounded shapes of fruits, pebbles, falling drops of water and lead, and even the heavenly bodies; the holding together of two smooth-faced solids; the existence of springs above sea level; why some things dissolve in various fluids and some do not. Hooke saw the solution to his problem as a way of perhaps gaining insight into all the phenomena of nature.

The basis of Hooke's explanation was the existence of what he called "congruity" in nature. By congruity Hooke meant adhesive force. Congruity is a property of bodies whereby they tend to stick together with other fluids or solids. Congruity included not only coming together but also staying together. Incongruity is a tendency to disunite. Congruity and in-

[9] Gunther, Vol. 7, 1/17/1678.

congruity can readily be seen operating in all sorts of common phenomena of nature. Some things, such as mercury, naturally tend to stick together, while other things, such as oil and water, naturally tend to stay apart. Another common example of the antipathy or incongruity among various things, claimed Hooke, is the tendency for air and water to separate. This can be seen in the tendency of a water globule to contract into its smallest possible space, a sphere, when in the air. The same can be seen in the case of spherical air bubbles under water.

Hooke's physical explanation of capillary action and related phenomena was that there is a greater sympathy or congruity between glass and liquid than between glass and air. The Curator thought that he had to establish the truth of two propositions in order to prove his case. One was that unequal air pressures inside the tubes will cause unequal heights in the fluid levels. The second was that such a situation actually existed in the case in hand.[10]

To prove the first proposition, he simply partially filled a hollow U-shaped tube with water and then proceeded to change the water levels in each arm by blowing and sucking on one of the open ends.

To prove the second, he devised a means to show that air pressure inside tubes will decrease along with the diameters. Hooke took a glass tube that extended down about three feet, curved around and up into a small bowl, the top of which contracted to a small opening, and successively fitted fine tubes of varying diameters into the opening on top of the bowl. In each case, the fine tube was then filled with water but the bowl was left empty. Water was then poured down the long arm of the tube until the pressure was sufficient to force an air bubble up the fine tube.

He observed that the larger the tube cemented into the bowl opening the more water that was needed to push an air bubble up through the tube thus forcing out the water held therein. Since greater pressure was needed, water has a greater tendency to adhere to the walls of the larger glass tubes than does the air bubble. Hence, air pressure within the fine tubes exhibiting capillary action must be less than the outside air pressure. The conclusion, states Hooke, necessarily follows; namely, "That this unequal pressure of the Air caused by its ingress into unequal holes, is a cause sufficient to produce this effect, without the help of any other concurrent; and therefore is probably the principal (if not the only) cause of these Phenomena." [11]

[10] M., p. 11.

[11] M., p. 21. It might be noted that Newton, as revealed in his second paper on light (1675), accepted Hooke's explanation as the true one.

DIAGRAM ILLUSTRATING HOOKE'S EXPERIMENT ON CAPILLARY ACTION

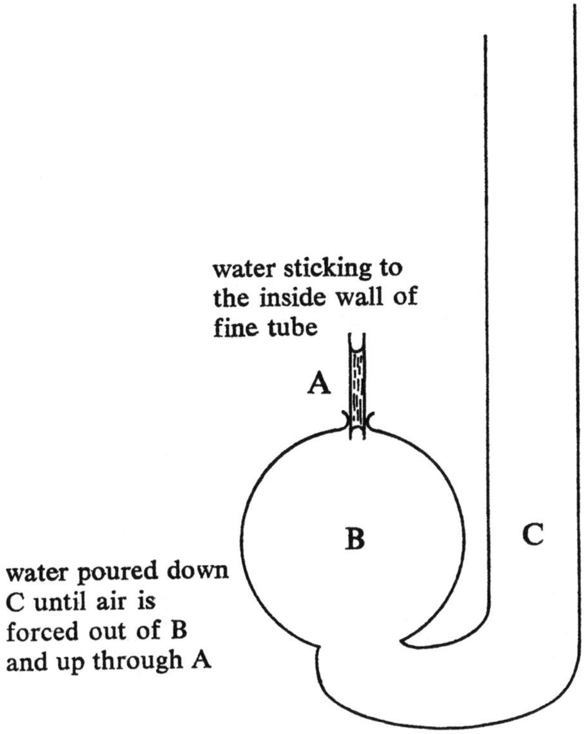


Today, capillarity is thought to be the resultant of four forces acting simultaneously at the junction of the liquid with a wall and vapor. Three of the forces are the surface tensions of the (1) solid-vapor boundary film, (2) the liquid-vapor film, and (3) the solid-liquid film. The fourth is the adhesive force between the small quantity of liquid at the junction and the wall. If the surface tension of (1) is greater than that of (3) the liquid will form a contact angle of less than 90 degrees and be seen to rise up the side of the wall. In a small cross section tube the liquid will rise until an equilibrium height is reached.

Hooke's error lay in too quickly assuming that his simple explanation accounted for all the facts. However, his approach was praiseworthy and his observations accurate. Capillarity, Hooke suggested, may account for the rising of fluids in wicks, sponges, plants, etc. Also, unequal air pressures most likely account for springs above sea level and may even explain the shapes of fruits, stones, and the heavenly bodies. We see here a man not wishing to unduly complicate nature. He believed that what he discovered from his experiment was sufficient to explain the phenomenon in question and perhaps many others. Justifiably, he felt no need to seek further and so he did not.

### *3.3 Fluid Pressure (1662).*

About the same time in his life that he was interested in capillary action Hooke performed two related experiments concerning fluid pressure. In January of 1663 he submitted a paper entitled "An account of some trials for the finding how much, ascending and descending bodies press upon the medium through which they pass: made before the Royal Society, Dec. 24, and Dec. 31, 1662." [12]

This short treatise contains a report on two experiments which Hooke had made on the effect, with respect to the resulting increase or decrease in fluid pressure, of a body passing through the fluid. These two experiments appear to represent an isolated bit of work in Hooke's repertoire of interests. One does not find anything quite like it either before or after.

The Curator was curious to know whether or not a body passing through a fluid medium would increase the weight of that medium. Would a descending object add or subtract weight from the fluid; if so, how much? Would an ascending body do either and, if so, to what extent? To solve his problem, Hooke designed two experiments, the first of which was subdivided into two parts.

The procedure for the first part of Hooke's first experiment involved hanging a two-foot long tube, closed at one end and filled with water, from one arm of a beam balance. Attached to the top of the tube, and projecting over the mouth of the tube, was a wire. Attached to the wire was a glass weight, submerged in the water, and suspended by a thread. The apparatus was then balanced.

For the second part of the first experiment, the Curator (using the same arrangement as before) ran a thread down to the bottom of the tube and then through a little loop atop a small weight resting at the bottom. Just after passing through the loop, there was attached to the thread an object that would float to the surface when released. The apparatus was then balanced.

The entire experiment consisted in cutting the thread in each situation and observing what happened. This was done; the results: nothing in

[12] See Gunther, Vol. 6, 12/24 and 31/1662.

either case. This is to be expected, as can be seen from the arrangement of the apparatus.

On December 31, Hooke tried a variation. Instead of hanging the glass ball from a wire attached to the tube, it was attached to a projection over the mouth of the tube which was independent of both the tube and the balance. After balancing the apparatus, the thread was cut as before. This time, and on several retrials, the balance moved down on the tube side. It does not appear that Hooke tried this variation with an ascending body.

What did Hooke conclude on the basis of his experiments? "These experiments," said Hooke, "seem to hint this axiom, that every body, whether ascending or descending in a fluid body does add so much weight or pressure to that fluid body, as its own weight amounts to, and not as much as the weight of so much of the fluid as is equal in bulk to what the moved body's amount to." [13]

It was well known in Hooke's day that an immersed body will be buoyed up with a force equal to the weight of the displaced fluid.[14] Hooke, however, found that in his experiments the added weight amounted to the weight of the submerged body and not to the weight of an equal volume of fluid. Hooke did not follow up this line of reasoning, however, probably realizing that his results did not contradict Archimedes. Since the scale was balanced to begin with, he expected the added weight to turn the scales by that amount and he was not disappointed. Since this was what he expected there was no need to continue his efforts in that regard. So he immediately turned his attention to other considerations.

These other considerations take the form of difficulties which Hooke felt prevented him from giving total assent to his conclusion. As part of the first difficulty, he states that the resistance of a medium to a body moving through it is directly proportional to the body's speed so that the lower, and hence more swiftly, a body descends, the greater should be the resistance.[15] It would seem, therefore, that as the body fell in the tube the pressure it was exerting against its container should have exceeded its own weight due to its increasing speed. This did not occur and Hooke does not attempt to say why.

[13] *Loc. cit.*

[14] See M. Clagett, *Archimedes in the Middle Ages* (Madison, Wisconsin, 1964) for a detailed account of the translation history of Archimedes. See also G. Sarton, *The Appreciation of Ancient and Medieval Science during the Renaissance* (New York, 1961), Lecture III, #3 on the importance of Archimedes with respect to the founders of modern mechanics.

[15] See G. Galilei, *Dialogues Concerning Two New Sciences* (tr. and ed. by H. Crew and A. de Salvio), First Day, Section 119.

DIAGRAMS ILLUSTRATING HOOKE'S EXPERIMENTS ON FLUID PRESSURE

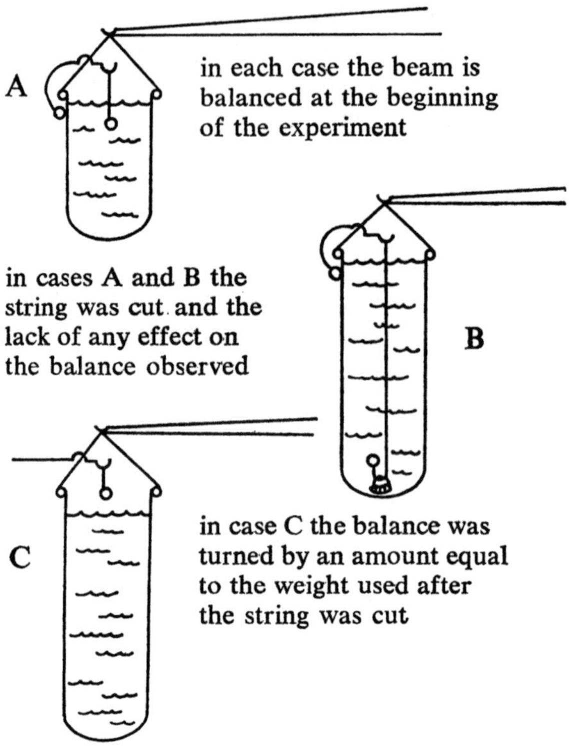


Hooke also expressed indecision as to exactly where the increased pressure, supplied by the weight moving through the fluid, should be exerted. Was it at the bottom, just under the weight, all over, etc.? If the weight of the body were sustained throughout the whole of a fluid, why then does a body falling through the air hit the ground so hard? Furthermore, it seemed to Hooke that a body ascending in a fluid contradicted the fact that the pressure of a fluid against the bottom of its container was directly proportional to the height of the fluid. As far as Hooke could tell, the floatable body rose no more quickly when near the top than when near the bottom whereas it should have done so if the fluid pressure decreased. That the pressure of a fluid was directly proportional to its height was well known in Hooke's day. Simon Stevin (1548-1620), who lived in Holland where such things were important when it came to building dykes, had already established the fact around the turn of the century.[16] In ad-

[16] See G. Sarton, "Simon Stevin of Bruges," *Isis*, Vol. 21 (1934), pp. 241-303, especially sections 13, 45-52.

dition, Hooke does not mention Pascal's law; namely, that a pressure applied at any point to a confined fluid is transmitted with equal intensity in all directions within the body of the fluid. It may be that he was aware of it but saw that he and Pascal were working with different experimental situations; namely, Pascal was working with a sealed-in fluid while he was not.

Hooke did not attempt to resolve these difficulties. Instead, he went on to draw three conclusions from his experimental results which he considered to be certainly true. These were that vapors from the earth press upon the earth with the same pressure whether ascending or descending, that the pressure on the sides of a vessel is reduced by opening a hole in its bottom and allowing the fluid to run out, and, as an application of conclusion two, it would seem that the pressure against the supports of a bridge is less when the water is allowed to run by more easily. The Curator does not seem to be the least bit aware of the possibility that the laws governing falling liquids might differ considerably from those governing liquids moving laterally. Strangely enough, however, he nevertheless hit upon a vague foreshadowing of what today is called Bernoulli's Principle, after Daniel Bernoulli (1700-1782); namely, that as the velocity of a fluid over a surface increases the pressure upon that surface decreases. Hooke was totally unaware of this principle in its generalized form. Rather, he seems merely to have envisioned bridges with thinner and more numerous supports which would withstand floods much better than the usual kind. Apparently, Hooke was happy to have learned something of practical significance from his work and does not seem to have ever troubled himself again about some of the unexpected results of his work.

### *3.4 The Springiness of Air (1660-1678).*

In contrast to the seeming fleetingness of his interest in experiments on capillary action and fluid pressure, the "springiness" of the air interested Hooke over a long period of time. Viewing Hooke's work as a whole, there appears to be a good reason for this prolonged interest. As we will see in detail in a later section on Hooke's Law, the springiness (elasticity or elater) of the air illustrated for Hooke one of the most fundamental laws of nature; namely, that there is a one to one correspondence between the motive power bringing about some change and the amount of change produced. Let it suffice for the present, however, to say that it may well have been the case that as Hooke's Law crystallized in Hooke's mind, his work on the springiness of the air became more important. This was not the case at the outset of Hooke's work, however. As with capillary action, the influence of Boyle was what originally moved Hooke to action. The springiness of the air was only one of the topics concerning the air that interested Boyle, under whose direct influence Hooke worked when he first began experimenting on the subject of the air. In fact, as will be noticed shortly, it was not the springiness of the air which primarily interested Hooke in the early 1660's. Rather, gaining knowledge about the springiness of the air was a means to obtaining an explanation of other things.

The places in Hooke's works wherein he discusses the springiness of the air are spread out over an eighteen year period. The first experiment on this topic related by Hooke took place on August 2, 1660.[17] Exactly one year later, according to Hooke, he again attempted the same experiment. He did, in fact, re-use his same carefully preserved equipment.[18] About this same time, the summer of 1661, Hooke was experimenting on the springiness of the air by another, different and simpler, type of experiment.[19] The problem, procedures, and results of these experiments are recorded in Hooke's *Micrographia.*

Other experiments of the same nature appear in the records of the Royal Society but not in the *Micrographia*. In December of 1662 Hooke made trial with alcohol, instead of air, and mercury as well as with regular air and mercury. These trials did not work out as well as the others, however.[20] In 1678, Hooke was again called upon by the Society to repeat his experiments in order to silence incredulous members who did not believe that to condense the air twice required twice the pressure, to do it three times took three times the pressure, etc., and inversely, the force of the spring of the air diminished in proportion to the air's expansion so that half the quantity had but half the strength.[21]

In outline, these experiments were all basically the same. Hooke was defending the same conclusion in 1678 as he was in 1660. There does not appear to have been any evolution in basic problems, methods or results over the years.

Hooke's problem here, as with capillary action, was twofold. His immediate task was to establish a hunch he had about 1660 concerning the relationship between the volume of air and its pressure. The exact origin of his hypothesis is not revealed to us by Hooke. The more remote purpose

[17] See M., pp. 222ff.
[18] See M., pp. 225ff.
[19] See *loc. cit.*
[20] See Gunther, Vol. 6, 12/10/1662.
[21] See Gunther, Vol. 7, 2/7 and 14/1678.

(but to Hooke's way of thinking, the primary problem to be solved) was the odd appearances of various things as seen through the atmosphere, such as the distorted shapes of the sun and moon seen on various occasions. Part of his explanation depended upon the existence of air far above the earth. As Hooke put it in his *Micrographia,* just before explaining his work on air pressure, it was an established fact, verified by use of the "Torricellian Experiment" at various heights, that air pressure varies with different heights. None of those who had performed the experiment, however, had given any kind of formulation of the relationships among volume, weight, height and pressure. The results of the experiment merely showed in a vague way that the higher one went up the less weight the air seemed to have. This implied that there was less air. But did the air end somewhere? Could it extend far enough to bend light rays far above the earth? Merely assuming that the air indefinitely thins out as one went up was not good enough for Hooke. He needed some experimental evidence to justify his view on at least the possibility of an indefinite extension of the air. Hooke's task then was to bring more precision into this area. Because his work on air pressure will be useful as part of his explanation of various anomalies, states Hooke, he will bring it into the discussion. Let us sample Hooke's simplified experiment designed "to find what degrees of force were requisite to compress, or condense, the Air into such or such a bulk." [22]

He accomplished his purpose by bending a tube, closed at its bottom end, into a J-shape. The long arm was fifty inches long and the short arm was seven inches long. He then poured mercury down the long end until it compressed a certain amount of air. The air was then compressed to half its bulk and the amount of mercury noted. This was continued over and over again, each time compressing the enclosed air to half its former volume. As shown in a table of figures, the pressure needed to compress the air (i.e., its Elater) varies inversely with its extension. "From which Experiments," concluded Hooke, "I think, we may safely conclude, that the Elater of the Air is reciprocal to its extension, or at least very neer. So that to apply it to our present purpose (which was indeed the chief cause of inventing these wayes of tryal) we will suppose a Cylinder indefinitely extended upwards." [23] Hooke then goes on to show that by substituting the notion of weight for pressure and height for volume in the "$PV = c$" rule one can effectively argue that the atmosphere fills the whole

[22] M., p. 225.
[23] M., p. 227.

solar system. A glance at the following formulas will indicate why such a substitution is possible.

$$P = \text{weight/area}$$
$$V \text{ (of a cylinder)} = \text{height x base area}$$

Since the areas are constant throughout Hooke's explanation, they can be ignored. This leaves him, when discussing the extent of the atmosphere, with "c = WH." Let us follow his reasoning more closely.

Hooke supposed that the air pressure on the surface of the earth was equal to thirty inches of mercury. The question now arises as to how far the air must extend in order to produce this pressure. Hooke explained that Boyle had already established the fact that the weight of quicksilver to air was 14,000 to one. And, supposing the parts of a cylinder of atmosphere to be of equal density going up into space, the air would extend upward for seven miles, assuming 5,000 feet to each mile. The simple mathematics of the situation would be to relate the relative weights of air and mercury, which are known, to their relative heights, one factor of which is known, namely, the height of mercury to be thirty inches. Solving for the unknown height of air, we find that 30 inches is to 420,000 inches or 35,000 feet (i.e., seven miles) as 1 is to 14,000. Now, says Hooke, let us *assume* that a cylinder of air is divided into a thousand parts each thirty-five feet long so the bottommost section presses with full force upon the earth. But, the section above the lowest would not press with full force upon the earth nor upon the lowest section. The weight of the above air will decrease as its volume decreases so each higher section will press with less and less force upon each lower section.

But now, using his work on the reciprocal proportion to be found between the weight and extension of the air, Hooke concludes that the successive cylinders need not be an actual thirty-five feet each but rather compressed or expanded in volume depending upon the weight of the above air. That is to say, the air thins out as one becomes further removed from the earth with the result that, as the pressure capable of being exerted by any particular section of air decreases, its height would increase. The *total* pressure, however, would remain constant, about thirty inches of mercury. For, he states, "as the pressure sustained by the 999th is to the pressure sustain'd by the first, so is the extension of the first to the extension of the 999th so that, from this hypothetical calculation, we shall find the Air to be indefinitely extended."[24] Learning this, interjects Hooke, "was indeed the chief cause of inventing these wayes of tryal."[25]

[24] *Loc. cit.*
[25] *Loc. cit.*

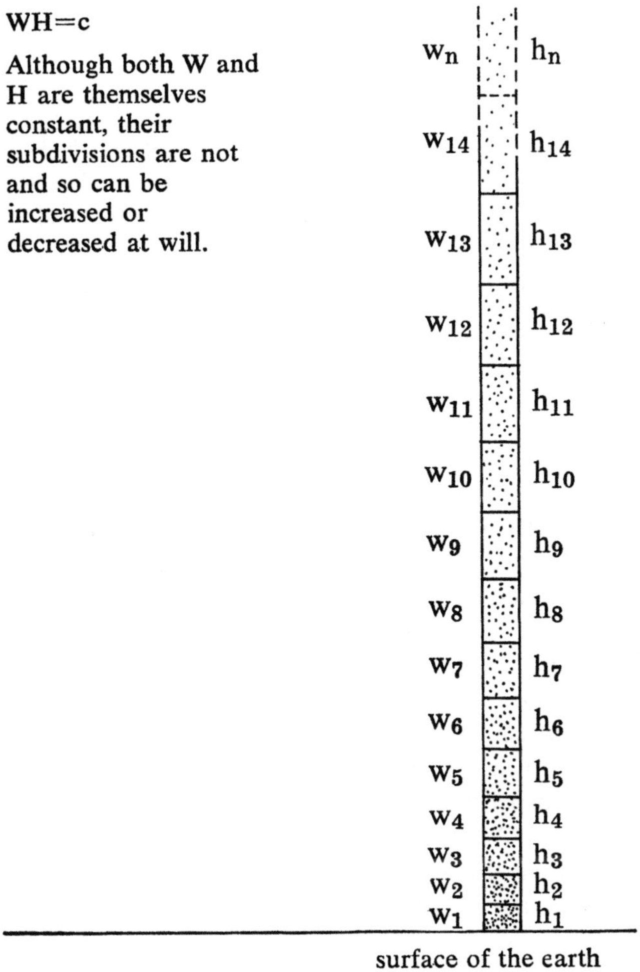


DIAGRAM ILLUSTRATING HOOKE'S ARGUMENT FOR THE VAST EXTENSION OF THE AIR

A more modern formulation of what Hooke said might read as follows. The pressure sustained by the 999th section of Hooke's imaginary cylinder would be equal to the weight of the 1000th section. The pressure on section one (the lowest) would equal the combined weights of sections 2 to 1000. Letting w stand for weight, e for extension or height, and the subscripts for the particular sections involved, we could write the following formula:

$$\frac{w_{1000}}{w_{2\text{-}1000}} = \frac{e_1}{e_{999}}$$

That is to say, if the weight of the 1000th section is very small, the height of the first section is very small and, if the weight of sections 2 – 1000 is relatively very great, the height of the 999th section is relatively very great. *A fortiori* the extension of the 1000th section will be very great indeed.

Hooke is ingenious here. Although, as he admits, he has not conclusively proven his case for the infinite extension of the air, he believes he has made it likely. He did this by taking liberties with the "PV = c" rule. By itself this rule (even in its atmospherically oriented form of "WH = c") could not yield what he wanted. He could not make W smaller in order to make H larger because W was itself an empirically observable constant (about 30 inches of mercury), while H also, whatever its actual value, would also be a constant. He could and did, however, break up W and H into many subdivisions each of which could be made numerically smaller or larger so as to successively increase the value of H. W would then become a combination of weights, e.g., $W = (w_1 + w_2 + w_3 + \ldots . w_{1000})$ and H a combination of heights, e.g., $H = (h_1 + h_2 + h_3 + \ldots . h_{1000})$.

The result of this division was a rather significant innovation in one's view of the air, namely, that the air at sea level is *compressed* air. Hooke realized that by stacking up cylinders of air, each with an equal and constant weight, one on top of another, the lower sections would be compressed thus causing the seven mile column of air to shrink. His task was to stretch it out again. He found that by diminishing the successively higher weights and increasing the successively higher heights it became a simple mathematical problem to stretch the column of air out to infinity. After summarizing his experimental results on the springiness of air by the simple inverse proportion "PV = c," he found it a simple task to apply his formula to the air and thereby to regard his view on the vast extension of the air as also experimentally well-founded. Once he had made it plausible, without doing violence to the empirical data, that the air extended a vast distance above the earth, he could then continue his reasoning in order to attain his ultimate end, namely, explaining certain unusual atmospheric phenomena in a simple, mechanistic way.

Next, since there is also no reason to suppose that there are any radical jumps in density between the various layers of air, as, for instance, there is between water and air, one can suppose that light rays are continually bent from their straight-line paths. This fact, in turn, can be used to explain all sorts of phenomena, such as the redness of the sun, the colors of distant objects on earth, and the varying shapes of objects seen through the atmosphere. Hooke listed a dozen or so items that might be causally explained by his view of the air. However, he warned, "these are but conjectures also, and must be determined by such kind of Observations as I have newly mentioned." [26]

At the conclusion of his report on his air experiments in the *Micrographia,* Hooke expressed the view that he had contributed something new to man's knowledge of the atmosphere. However, it was not the common textbook version of the discovery of Boyle's Law that was emphasized. "For since (as I hope)," said Hooke, "I have here shown the Air to be quite otherwise then has been hitherto suppos'd, by manifesting it to be, both of a vast, at least an uncertain, height, and of an unconstant and irregular density; it must necessarily follow, that its inflection must be varied accordingly." [27]

It must be remembered that Hooke lived in an age when many things we regard as commonplace and ordinary were regarded as mysterious. Bacon had taught Hooke to go to nature, probe around, and not be satisfied with complicated or mysterious answers to problems concerning nature. Now, granted that some problem concerning nature was important in the Baconian sense of greatly affecting human existence, and granted that one had the means for probing for an answer, Hooke would say, as we saw in Chapter II, that the proper approach was to attempt unifying all the available evidence with as few explanatory principles as possible. This was Hooke's attitude toward the atmosphere. And, as a result of his investigations, he firmly believed that he had solved not only a major set of problems touching the air but, by regarding the atmosphere as a unifying principle, he could also resolve other anomalies witnessed in nature. In an age which had not yet given up occult powers and the belief in the dichotomy of the universe with respect to the areas above and below the moon, this was significant. Hooke was claiming credit for showing that what had previously been considered diverse phenomena were actually all different manifestations of effects produced by the atmosphere. Since the air varies in composition and density, all sorts of bendings of light rays can take place, thus accounting for all sorts of anomalies in appearances. His direct achievement was to use what he had verified about the relationship between the air's condensation or rarefaction and its accompanying pressure, i.e., its springiness, to show how certain phenomena, previously thought to be the

[26] M., p. 240.

[27] M., p. 236. Mariotte and Halley later estimated the air's extension as 35 and 45 miles respectively. See A. Wolf, *A History of Science, Technology and Philosophy in the Sixteenth and Seventeenth Centuries* (2nd ed., New York, 1959), Vol. I, pp. 314-316.

results of different and mysterious causes, could be simply and adequately explained. Considering the era in which he lived, Hooke's approach to nature should not be underrated. The over-all attitude exhibited by Hooke goes a long way in marking him as one of the more advanced thinkers of his time. His indirect achievement, however, was the verification of Boyle's Law.

### *3.5 "Boyle's Law."*

From our modern point of view, it is the so-called Boyle's Law (Mariotte's Law on the Continent), namely, gas pressure times volume equals a constant, that is significant.[28] However, from Hooke's point of view, this was not the case. As we saw in the last section, it was not Hooke's primary intention to demonstrate what today is generally called Boyle's Law. Rather, Boyle's Law was a step in the process of explaining various optical anomalies seen in the atmosphere. Nevertheless, the origin of Boyle's Law appears to be of continuing interest to twentieth century historians of science.[29]

Yet the question remains: Is Boyle's Law truly Boyle's? The historical records allow us to give a fairly clear answer to the question. There is evidence that Hooke had much more to do with Boyle's Law than he is generally given credit for. He was, in fact, the first to verify it if not the first to enunciate it. In 1660 Boyle published his *New Experiments Physico-mechanical touching the Spring of the Air and its Effects, Made for the most part in a New Pneumatical Engine.* In this work, the relationship between the pressure and volume of air is not explicitly stated quantitatively. Boyle's book was criticized shortly after its publication by F. Linus (1595-1675), a professor of physics at the University of Liege. In his *Tractatus de Corporum Inseparabilitate,* Linus fought desperately against having to admit the existence of a vacuum, i.e., a space of non-being. Linus claimed the existence of Funiculus, a very thin substance pervading space which caused bodies to act against their natures, e.g., mercury rising in a tube instead of flowing down. Hobbes, also, joined in the criticism. Hobbes' approach was to ridicule and laugh at Boyle and his friends for experimenting.

Boyle's answer came in 1662 in the form of an appendage, entitled *A*

[28] The Frenchman Edme Mariotte came upon the law independently in 1676.

[29] See, for example, E. Andrade, "Robert Hooke," *Proceedings of the Royal Society of London,* Series B, Vol. 137 (1950), pp. 153-184.

*Defence of the Doctrine touching the Spring and Weight of the Air,* to his 1660 work. In this later work, Boyle's Law is explicitly stated. Boyle, however, did not claim to be its originator. On the contrary, he mentioned to his readers that his assistant Hooke had known of the relationship in question about 1660 and possibly before. Boyle, in order to be fair, relates how a certain "ingenious gentleman Mr. Richard Townley" had told him that he was working on the problem of the relationship between air pressure and volume. Towneley, however, as far as Boyle knew, had not actually verified the "PV = c" rule. And, since Boyle had no way of contacting Towneley, knowing when if ever Towneley would publish his views, or even if Towneley had the means to carry out experiments, Boyle decided to "present the reader with that which follow, wherein I had the assistance of the same person, that I took notice of in the former chapter, as having written something about rarefaction." Boyle relates further how Hooke, upon hearing Boyle mention Towneley's hypothesis, said that he had the year before experimented on that very subject with positive results. Boyle also mentions that Lord Brouncker, too, was doing some work in that area but had not achieved anything conclusive.[30]

Since the law was published in a book under Boyle's name, the law became generally known as Boyle's. However, as we have seen, within the very same book, Boyle disclaims being its discoverer and does instead credit "the same person, that I took notice of in the former chapter," who was indeed none other than his assistant Hooke, with both thinking upon and verifying the hypothesis in question. A "Boyle side" to the question of origination is, therefore, nonexistent, as Boyle himself testifies. Also, in his life of Boyle, More expresses the view that Boyle's Law was actually more Hooke's than Boyle's.[31] The experiments involved in its verification

[30] See Boyle, Vol. I, p. 102. It might be added in passing that it was probably Hooke who wrote the attack on Linus' "Aristotle's Wheel" argument against the existence of a vacuum which concludes Boyle's *Defence.*

[31] See L. T. More, *op. cit.,* pp. 94-96. But what of a possible Towneley claim to priority? In his 1661 retrial of his 1660 experiment on air springiness, Hooke said he had forgotten much of his earlier work on air by then and so resolved to redo his experiment. This time, he added, he intended to take into consideration Towneley's (spelled Townly by Hooke) suggestions and arrive at greater exactitude. Hooke himself is not much help with respect to Towneley. He simply states that in his retrials, "I did not exactly follow the method that I had used at first; but, having lately heard of Mr. Townly's Hypothesis, I shap'd my course in such sort, as would be most convenient for the examination of that Hypothesis." (M., p. 225.) The results, claimed Hooke, were basically the same as in his 1660 experiments.

According to the *Dictionary of National Biography,* Richard Towneley of Towneley Hall, Lancaster, was a country gentleman noted for interests in ancient literature and the new sciences. His son, Christopher Towneley (1604-1674), the only one explicitly

were, of course, the same ones referred to in our discussion of Hooke's work on the relationship between springiness, density, or volume of the air and its pressure.

For his part, as expressed in the *Micrographia*, Hooke's general attitude with respect to the law in question seems clear. Nowhere does he credit Boyle with its original enunciation and verification which he certainly would have, considering their close and friendly relationship, if it were true. Rather, it was he who verified the law by experimentation even though one or two others may have been the first to suggest the law. Others, however, never confirmed it; Hooke did. We note, then, a fundamental agreement between Boyle and Hooke on the issue. Henceforth, therefore, let us give Hooke credit for at least scientifically establishing the law in question.

### *3.6 The Debates of 1678.*

About fifteen years after Hooke did his main work on fluid pressure, his theory on the cause of atmospheric pressure came before the public in a series of debates on the subject carried on within the Royal Society. Toward the end of 1677, Hooke expressed the view that differences in air pressure were due to differences in the amount of vapors in the air. This, he claimed, agreed with the principle that the weight of a fluid upon equal areas of a container is always a function of the weight times the height of the fluid.[32] A week later, in a debate with Dr. Croone, Hooke maintained that the shape of the container was irrelevant; the pressure depended only upon the weight and height of the contained fluid.[33] Two weeks later, 3

listed in the *Dictionary of National Biography*, had similar interests which he followed in the form of correspondence with noted men of his time on various subjects. A younger relative of his, also named Richard, the Towneley referred to by Hooke, contributed several articles to the *Philosophical Transactions* and is mentioned several times in Hooke's *Diary*. (See G. Keynes, *A Bibliography of Dr. Robert Hooke*, pp. 6, 8.) Other than this, practically nothing in known concerning the Towneleys' relations with their contemporaries. Newton, on page seven of his handwritten notes on the *Micrographia*, notes that "Mr. Townlys Hypothesis is the dimension (or expansion) of the aire is reciprocall proportiona to its spring (or force required to compresse it). By Mr. Hookes Experience . . ." Newton then copied out the table of figures given on page 226 of the *Micrographia*. (See *ibid.*, p. 107.) Also, in Newton's *Principia*, no. 68, p. 609 in the Cajori edition we read: "And having this ratio, we may compute the rarity of the air . . . supposing the expansion thereof to be inversely proportional to its compression; and this proposition has been proved by the experiments of Hooke and others." Marie Boas states flatly that what later became known as Boyle's Law was discovered independently by Towneley and Hooke. (See M. Boas, *op. cit.*, p. 44.)

[32] See Gunther, Vol. 7, 12/13/1677.

[33] See Gunther, Vol. 7, 12/20/1677.

January 1678, the topic was again raised. Was not clear air as heavy as foggy air, asked Henshaw, the Vice President. Hooke retorted that the air will remain clear as long as the ether perfectly dissolves whatever is in it, like salt or sugar in water. Fogginess results from the separation of the ether's contents from the ether, as, for instance, when something does not dissolve in water, but floats about in it clouding it up. The weight of these separated substances does not change, however, because they still remain suspended in the ether.[34] Therefore, in answer to Henshaw's query, clear air can be as heavy as foggy air, just as water will be the same weight before a substance dissolves in it (thus giving it a cloudy appearance) as it will be after the substance dissolves in it (thus making the water appear clear).

At the next Society meeting Hooke again defended his position. Ether, said Hooke, penetrates all, even glass, which acts as a strainer separating out those things that may be dissolved in it. Henshaw asked about the weight of air in damp weather. Hooke replied that the mere dryness or wetness of the ether had no effect upon the air's weight. The only important factor was the amount of exhalations suspended in a given volume of ether.[35] The following week saw the matter discussed once more. Hooke mentioned his barometer observations in order to show how they supported his view. He also noted how air must be like a fluid considering the way clouds floated on it.[36] By this time the members apparently felt they had exhausted the subject and the matter was dropped for a while.

Approximately a year later, however, the issue was again mentioned. Why was the barometer lower in rainy weather than in fair weather, some of the members wanted to know. Because, explained Hooke, in wet weather the ether could not take up as much of the parts of other bodies as it could in a dry weather.[37] And so the issue rested. Thus, after the public debates which revealed to Hooke that there were no telling objections againt his view, he was satisfied that this view on the nature of air (i.e., air is a collection of earthly exhalations suspended in the ether) accounted very well for all the facts of experience.

In a way this is somewhat curious. One would not expect this "on faith" acceptance of the mixture theory of air of an adherent of experimental philosophy. But in this case, as with Descartes' general scheme of things, as we will see, this is exactly what happened. The explanation,

[34] See Gunther, Vol. 7, 1/3/1678.
[35] See Gunther, Vol. 7, 1/10/1678.
[36] See Gunther, Vol. 7, 1/17/1678.
[37] See Gunther, Vol. 7, 2/6/1679.

of course, is that Hooke was neither a perfect scientist nor a perfect philosopher. In contrast to something such as the explanation of gravitation, over which there was much controversy among his associates, the mixture theory was the generally accepted view among the people he respected. As far as we know, he never set out to test the mixture theory itself. No doubt, if, in his other enterprises, he had found some evidence that contradicted the theory, he would have pursued the issue further. But he did not find such evidence and died believing Boyle correct.

CHAPTER IV

# THE MECHANICS OF TERRESTRIAL LOCAL MOTIONS

## *4.1 The Legacy of Descartes.*

Our task in this section is to complete the background needed to make Hooke's work in mechanics intelligible. This means we must make explicit what previously has been hinted at in passing. We have seen Hooke as a convinced Baconian in methodology; we must now see him as a convinced Cartesian in his conception of the physical world. To fulfill our aim, those aspects of Descartes pertinent to Hooke will be set out. Descartes' views will be followed by Hooke's statements on the same topic thus making clear the doctrinal correspondence between the Frenchman and the Englishman. We will also again see the intermediary role played by Boyle, this time between Descartes and Hooke.

According to Waller, Descartes was the only major seventeenth century scientist of whom Hooke had a first-hand knowledge as a young man. This undoubtedly included Descartes' *Principles of Philosophy* in which he discusses *ex professo* his philosophy of nature. This work appeared in Amsterdam in 1644. Its four books dealt respectively with the principles of human knowledge, the principles of material things in general, the visible world, and the earth. According to Descartes' conception of our universe, the only two principles needed to explain all natural phenomena are matter and motion. God created matter and set it in motion. He also conserves them in existence and, because of His trustworthiness and immutability, will not allow their total quantity to be either augmented or diminished.

But what are matter and motion? Matter for Descartes is simply that which is extended in three dimensions. As he states, an "extended object is called by us either body or matter." This can be known from the fact that it is possible to abstract all other attributes from our concept of matter except that of extension. So it is that "the nature of body consists not in weight, nor in hardness, nor color and so on, but in extension alone." [1]

Also, matter is continuous; the universe is a *plenum* with every part affecting every other part throughout the whole expanse of the universe. "There is therefore," Descartes tells us, "but one matter in the whole universe, and we know this by the simple fact of its being extended." [2]

Once matter had been set in a spinning motion by God, innumerable whirl-pool actions among the different parts of matter took place. As time went on, much of the matter was ground up into finer and finer pieces. These formed the luminous heavenly bodies. Other particles, larger than the first type, but yet very small, round, smooth, and imperceptible constitute the ether. Other pieces of matter, gross and opaque, form the comets and planets. The sun is at the center of one vortex, while the planets (each in its own vortex), are bodies carried around in the current.[3]

In such a system there was no need for any explanations other than mechanical processes operating according to fixed laws. By means of various simple experiments, Descartes was able to illustrate his point. For instance, in a swiftly rotating bowl, filled with fine lead shot and lumps of wood rotating on top, the wood can be seen to move toward the center. A similar situation exists in the case of bodies "floating" in the ether around the earth. As the ether particles swirl around, they are constantly striving away from their vortex's center due to centrifugal force. As these fine spherical particles move around and press outward, the grosser, ponderable particles will "gravitate" to the center.

The Curator chose to follow Descartes' doctrine on matter and motion. This does not mean that the particular mechanisms he chose to explain various phenomena mimicked those elaborated by Descartes. What it does mean is that Hooke's approach to motion and force was a mechanistic one rather than a psychic, magnetic, or agnostic one, even if Hooke did not employ etherial vortices.[4]

Hooke himself never wrote, as far as we know, a separate treatise on matter and motion. His interest in motion was always united with some

[1] P.P., II, 4.

[2] P.P., II, 23.

[3] See P.P., III, 45, 52. Descartes did not mean to say that our cosmos evolved out of undifferentiated matter. He believed it was created more or less as it is now. However, he thought that an evolutionary view would help us better understand our world.

[4] See P.W., p. 177. See M. Jammer, *Concepts of Force* (New York, 1962), chs. 5 and 6 for a study of the other views of force.

particular object or group of objects such as the motion of muscles or the air. He did, however, on one occasion, treat of matter and motion as a separate topic rather than regarding them as elements in other categories. This occured in his 1682 treatise on comets and gravity, "A Discourse of the Nature of Comets."

In the course of his discussion the Curator of Experiments paused in order to define the terms which he was using. The whole of reality that in any way affects our senses, he claimed, is composed of body and motion. Body is that reality which has extension every way. Body is a positive and immutable reality. It is not immutable with respect to its shape or figure, however, but only with respect to the over-all amount in the universe. It is not figure that makes a body. A quart of water, for instance, still remains a quart quantitatively speaking regardless of the shape of the container. The "anatomists," i.e., the atomists of old, who claimed that body was equal to little lumps of matter with definite shapes and sizes, were totally misled, claimed Hooke.

Hooke likened body to a female or mother principle. Of itself, it is wholly inactive and without form. In order to be determined, it must be impregnated by a male or "Spiritus" principle, i.e., motion. Of itself, matter or body is uniform, homogeneous, and essentially immutable. It cannot, Hooke asserted, be altered in total quantity by either condensation or rarefaction. Also, the universe is a *plenum* in which there exists among all sensible bodies the fine, material, insensible ether.[5] The ether, as we have seen to some extent and will see better shortly, was needed to account for various natural phenomena which Hooke felt could not be otherwise explained.

Turning now to the subject of motion, we find that for Descartes, motion "in the vulgar sense is simply the activity by which a body travels from one place to another." [6] In an aside to this passage Descartes adds: "Local motion, that is; there is no other sort I can think of and I see no reason to imagine any other to exist in nature." [7] Furthermore, he insists that motion is not an independently existing reality as is matter. Rather, it is something that happens to matter; "it is a mode of the mobile thing and not a substance, just as figure is a mode of a figured thing, and repose of that which is at rest." [8]

When discussing motion, Descartes distinguishes between the first cause of all motion, God, and the laws of motion. The latter he calls the three

[5] See P.W., p. 172.
[6] P.P., II, 24.
[7] *Loc. cit.*
[8] P.P., II, 25.

"laws of nature." The first two laws taken together comprise the law of inertia. Every moving body insofar as it can tends to continue in motion.[9] Moreover, such continued motion will be in a straight line.[10] The third law states that if such a moving body strikes another body that is relatively stationary it will either not move the struck body and be deflected (when the resistance of the struck body is such that it cannot be moved at all) or it will move the struck body in such a way that the degree of motion lost by the striking body and gained by the struck body will equal a constant.[11] With respect to his third natural law, Descartes points out that the relative forces of impact and consequent directions would be very easy to calculate if one were dealing with perfectly solid bodies and if there were only two such bodies involved in any one interaction.[12] However, he also points out, this is an ideal situation which the philosopher of nature will never find on earth.[13]

We see quite a revolution in Descartes' view on motion. For centuries it was thought that rest was a natural state requiring no cause, while changes of all types, including local motions, did require some cause. This situation has been nicely summarized by Koyré. When speaking of Newton's disrespect for Descartes, Koyré points out that Newton did not "mention that it was Descartes' formulation of the principle of inertia, which placed motion and rest on the same ontological level, that inspired his own." [14] For Descartes inertial motion is a state of being. The expression status or state of motion implied for the Frenchman and those who followed him that "motion is not, as had been believed for about two thousand years – since Aristotle – a process of change, in contradistinction to rest, which is truly a *status*, but is also a state, that is, something that no more implies change than does rest." It is precisely because inertial motion, like rest, is a state, continues Koyré, "that motion is able to conserve itself and that bodies can persevere in motion without needing any force or cause that would move them, exactly as they persist at rest." [15]

Hooke himself has very little to say on the subject of motion. Where he does mention the topic explicitly, as in his *Lectures De Potentia Restitutiva, or of Spring* (1678) and the discourse on comets and gravity, it is clear that he agrees perfectly with Descartes on the basic definition

[9] See P.P., II, 37.
[10] See P.P., II, 39.
[11] See P.P., II, 40.
[12] See P.P., II, 45.
[13] See P.P., II, 53.
[14] A. Koyré, *Newtonian Studies* (Cambridge, Mass., 1965), p. 65.
[15] *Ibid.*, pp. 66-67.

of motion. For Hooke, motion is basically local motion. It is nothing more or less than the relocation of bodies with respect to one another or of the parts of bodies with respect to one another. As Hooke states, motion is "nothing but an Alteration, or Power of Alteration, of the Minims of a Whole, in respect to one another, which Power may be increased or diminished in any assignable Quantity." [16] To see what Hooke has in mind one must imagine the smaller parts of one body or the various individual bodies within a group of bodies changing position relative to one another. The speed with which the parts of the whole change position can be increased or decreased from zero to some very high limit. If the parts are not presently actually changing position relative to one another, there is at least the power or potentiality for change present. If the parts are actually changing position relative to one another, there exists the possibility for reducing or increasing the speeds of the various local motions involved. This view of motion was entirely in keeping with Hooke's general mechanistic outlook.

It should be made clear that this basic agreement with Descartes on motion is only with respect to the basic definition of motion, the genus of motion, if one may so speak. It is not the case that Hooke agreed with Descartes on what particular type of local motion was responsible for some particular phenomenon. That is, there was disagreement over the species of motion. As we will see, a good deal of Hooke's work centered on vibratory motions (e.g., light and gravity) which are propagated in wave-like fashions. These specific types of local motions used to explain various phenomena were not Cartesian. Descartes sought to explain light and gravity, for instance, by other types of local motions than those used by Hooke. These differences, however, were quarrels within the same family. In some other family, such as the Aristotelian, for instance, local motion might be regarded as a subdivision of some broader definition of change. But the

[16] P.W., p. 172. Throughout his works Hooke's identification of motion with local motion is more implicit than explicit. This doctrine is always there under the surface and occasionally we see it break through. A good example of this can be found in the course of Hooke's attempt to interpret *Genesis* in terms of his mechanistic framework. He states: "For, as I shall after wards shew, it must in this Place be so taken, and not only for the Substance Water, but that God joining the second Power *Motion*, and impregnating the *Mater* with it, made every *Minim* of it to move with infinite Varieties of Motions: And so all those which before lay still, and changed not their Places and Positions in respect of one another, and were therefore called Earth, are now by this Insufflation, Incubation, or acting of the Spirit of God upon the Superficies, become a perfect Fluid, or a Water. No two contiguous *Minims* yet agree in Unity or Uniformity of Motion. And hitherto seems to be the History of the Creation, or making of the two first Powers, *Matter* and *Motion, Body* and *Spirit*, or *Matter* and *Form*." (P.W., pp. 174-175.)

Cartesian family, of which Hooke was a part, regarded local motion as basic and explained other changes in terms of the mechanical relocation of parts.

The Curator was also in agreement with Descartes concerning the latter's principle of inertia. Galileo had come close to grasping the concept of inertia but had failed to do so because of his ties to the ancient notion of a closed, finite, spherical universe. In such a world, if motion did continue indefinitely it would have to ultimately be in a circle rather than in a straight line. Descartes' world, however, was infinite in all directions. The concentric spheres are gone and all horizons are eliminated. In such a universe there is no reason to suppose that an unhindered body in motion will ever curve. Hooke also lived in such a world. As we have seen in his speculations on the extension of the air and as can also be seen in his 1670 report on his attempt to prove the annual motion of the earth, Hooke showed no hesitation in affirming the limitlessness of the universe. Given an infinite universe Hooke, as Descartes, found it reasonable to postulate inertial motion in a right line. Provided there are no obstacles, then, a body set in motion will continue indefinitely with rectilinear motion; it enters a certain state of being. Non-inertial motion, however, loses its position as a state in Descartes' eyes. Curved local motion, for example, does require a cause or causes. Hooke agreed with Descartes in these matters. What Hooke thought these forces were will be seen in the next chapter.

He goes on to say, as Descartes had said, that one must look to God as the ultimate cause of body and motion and as the ultimate cause, due to God's immutability, of the conservation of both body and motion. As Hooke explains, both matter and motion are the immediate products of God's will. Nothing can destroy them except God himself. What is more, God acted in a rational fashion assigning definite, regular, and predictable motions to bodies, which regularity can be discovered by human scrutiny. Discovering God's rules is, in fact, says Hooke, "the true end of the Science of Physicks, or of Natural and Experimental Philosophy." [17]

Precisely why Hooke accepted Descartes' doctrine of matter in motion (as he did the mixture theory of air), in what may appear to some today as an uncritical acceptance, is hard to say. As far as we know he never actually attempted testing the matter in motion doctrine or Descartes' three basic laws of motion. Perhaps it is more a question for sociology or psychology rather than the history and philosophy of science. We have similar phenomena today. For instance, the acceptance of organic evolution and materialistic mechanism, even on the human level, in biology. Hooke was

[17] P.W., p. 172.

a critical thinker and an avid experimenter, but he did not question everything. Certainly, if Hooke had discovered on his own various things which contradicted the basic Cartesian philosophy, he would have searched for another. However, throughout his life, Hooke found that whatever he discovered or whatever explanatory theories he might propose could all use the Cartesian philosophy, whether exactly as proposed by Descartes or with some modification, as a backdrop. It was simple and adequate. Moreover (and perhaps of most importance – will we ever know?), it was favored by his teacher and friend Boyle, as well as by most other outstanding scientists of his day.

The situation might appear less obscure if we take into consideration what Emile Meyerson had to say in his *Identity and Reality*. According to Meyerson, all of natural science in its theoretical aspects is a search for identities; an attempt to demonstrate that seemingly great diversities are really but manifestations of more profound unities. Hence the universal tendency toward conservation or constancy laws and the popularity of mechanistic (especially atomic) theories among philosophers and scientists. The ultimate reason for this search for unity among diversity, says Meyerson, is the very construction of the human mind itself.

With respect to the principle of inertia (i.e., the conservation of velocity), for example, Descartes' argument in its favor "was sufficient to obtain the agreement of contemporaries: the principle of inertia, almost immediately dominated science, and this in spite of the fact that people had been accustomed for so many centuries to the antagonistic doctrines of Aristotle. Some have been astonished by this – wrongly, we think."[18] The acceptance of inertial motion as basic, as persisting throughout changes, as a state of matter underlying observed phenomena was merely a continuation of the progress of human thought, contends Meyerson. Due to the omnipresent influence of the principle of causality in human thinking, reasons Meyerson, our minds show an invincible tendency to maintain the identity of motion in time, i.e., to conserve it. "Here, it seems, is the true foundation of the principle."[19]

In order to explain how it was possible to have bodies affect one another over great distances Descartes thought it necessary to postulate the ether, which is divided up into many vortices or whirlpools. Descartes thought our earth was at the center of one such vortex; the sun at the center of another. The celestial bodies, under centripetal force, tend to collect at

[18] E. Meyerson, *Identity and Reality* (London, 1930), p. 145.
[19] *Ibid.*, p. 146.

the center because of a floating effect they undergo in the light, buoyant ether, under centrifugal force, as the ether swirls around.[20]

Descartes' basic doctrine on the existence and motion of the ether was adopted by Robert Boyle. With respect to the ether, Boyle opined that all of interstellar space was one vast ocean of ether wherein the "luminous globes" swim "like fishes."[21]

As one would expect, Boyle, in the role of Hooke's teacher and friend, exercised a great influence over his disciple's views. This fact, taken in conjunction with the general intellectual atmosphere of his time with respect to the ether and air, as well as the fact, related by Waller in his biography of Hooke, that Hooke read Descartes directly, makes Hooke's own position on the existence of ether quite understandable.

That Hooke firmly believed in an all-pervasive etherial substance cannot be denied. On one occasion, in 1678, when the Vice President of the Royal Society objected to some of Hooke's views on the cause of atmospheric pressure, Hooke unequivocally stated that he could prove its existence and properties beyond reasonable doubt.[22] For Hooke, the ether was a necessary adjunct to his theories on light, magnetism, gravity and air. True, it could not be seen but nevertheless it had to exist in order to make intelligible that which was observed. For Hooke, there was no such thing as action at a distance, as we will see.

Granted that Hooke believed in the existence of the ether and knew it to be but a subtle form of matter, the problem he then faced was whether or not it moved as Descartes said. Early in his career, Hooke seems to have gone along with the Cartesian notion of a moving ether. The ether concept was needed to explain the apparently circular motions of the heavenly bodies. The ether is moving *per se;* the planets are carried about *per accidens*. This can be seen in one early experiment performed by Hooke.

In 1662, while experimenting on the relative densities of hot and cold water, Hooke noted that the less dense hot water would rise to the top of the more dense cooler water. He thought this to be a minor discovery but, nevertheless, potentially useful in various ways. One use might be for the purpose of "conjecture," as he says. For, it may be that the "vast space of the vortex of the sun or the heavens" is filled with a moving fluid of different densities in which bodies float at different distances from the sun.[23]

[20] See P.P., II, 47ff.
[21] Boyle, Vol. 3, p. 706.
[22] Gunther, Vol. 7, 1/3/1678.
[23] Gunther, Vol. 6, 12/31/1662.

The fluid near the sun would be hot and thin, while farther away it would be cooler and denser. Also, he suggested, the planets may be hollow like glass balls and "be in an aequipondium to the ambient fluid." [24]

Although Hooke did not give up his belief in the existence of the ether, he did modify his earlier view by giving up the motion of a swirling ether. This was done primarily to account for his own theory of gravitation; a theory he considered more satisfactory than that of Descartes' because of its simplicity and conformity with experience.[25] We will discuss his gravitational theory in detail later.

This modification can be seen in two places in Hooke's work, once earlier [26] and once later [27] in his career. In the earlier passage this change of attitude is only implied while in the latter it is stated explicitly. We will therefore review both.

In the course of a discussion on universal standards of measure, Hooke suggested a short pendulum as an international time keeper. However, he was quick to point out the shortcomings of his suggestion. One shortcoming was that one could not be sure its motion would remain constant just as one could not be sure that the motions of the heavenly bodies had always been constant. For, Hooke continues, as one can observe when one spins a top or wheel here on earth, it soon slows down due to an impeding medium, the air. Likewise, the heavenly bodies, originally spun by God, are probably moving much more slowly today than they were in the first age of the world before the Flood. He adds that this may also explain those Bible passages relating the great age of the Patriarchs. Actually, they lived no longer than we, "though they might see and number ten times more revolutions of the sun and heavens." [28] As this passage implies, the medium remains still while the objects circulate through the medium.

The ether's lack of swirling motion is later stated explicity by the Curator.[29] In a discourse read before the Society in June of 1685, Hooke put forward his explanation for a strange light, periodically observed by various people, in the northern skies. Hooke thought the light to be caused by an extra abundance of effluvia from the earth's atmosphere about the time of its perihelion on 16 or 17 December. These steams from the earth are left behind in the ether due to the swift motion of the earth. Later, during aphelion, they are again picked up by the earth. In the reproduction

[24] *Loc. cit.* See also 10/29/1668.
[25] See Gunther, Vol. 7, 12/20/1682.
[26] See Gunther, Vol. 6, 12/14/1664. See also M., pp. 13-14.
[27] See P.W., pp. 196-197.
[28] See Gunther, Vol. 6, 12/14/1664.
[29] See P.W., pp. 196-197.

of Hooke's treatise in the *Posthumous Works* there are accompanying diagrams, the second of which depicts a plane semicircle on which is drawn the ball of the earth surrounded by dots representing the earth's effluvia. As an explanation of his diagram, Hooke explains that it represents the earth's path through the ether "which I suppose altogether stagnant, and not moving round with it in a Vortice, as Descartes supposed; but quiescent, according to the Theory of Celestial Motions which I long since have explained and shewn to this Society." [30] The theory referred to was contained in his 1666 paper on curved motion, to be discussed later.

Hooke even goes so far as to include his notion of ether in his digression on the meaning of *Genesis*. The whole universe consists of matter, motion, and the ether, which is nothing more than refined matter. The first thing God did was to make matter. Next, God put matter in motion. Along with motion came the two great laws of motion which constitute the form and order of the world; namely, light and gravity, both of which, claimed Hooke, could be explained in a purely mechanistic way. To Hooke, God's saying that there should be a firmament divided from the waters meant the introduction of the phenomenon of gravity. At this point, the heavenly bodies were formed in the ocean of ether.[31]

Hooke does not contradict what Descartes had said before him. With respect to celestial mechanics, however, Hooke was very much dissatisfied with Descartes' explanations. Both men were Copernicans but they differed as to the real, physical causes, accounting for the various movements involved in the heliocentric theory. The major difference between the two men was that Descartes affirmed and Hooke denied the existence of heavenly vortices. Hooke did so for what was to him the very best of reasons; namely, they contradicted experience.

Although the existence of comets which cut a path across the swirling vortices did not constitute for Hooke a major objection against Descartes, there were two other facts which did. One was the shape of the earth while the other was the direction of falling bodies. Everyone knew that the earth was spherical in shape. Hooke himself knew it so well that he nowhere attempts to defend the notion that the earth is roughly spherical. But how could such a thing be if Descartes were right? If Descartes were right, the earth would be a cylinder in shape. For, as the giant vortex in which the earth was supposedly centered swirled around, the grosser corpuscles would "float" toward the center which would not be a point but a line or axis of rotation. Also, the existence of heavenly vortices would

[30] *Loc. cit.*
[31] See P.W., p. 175.

DIAGRAMS ILLUSTRATING HOOKE'S OBJECTIONS TO DESCARTES' VORTEX THEORY

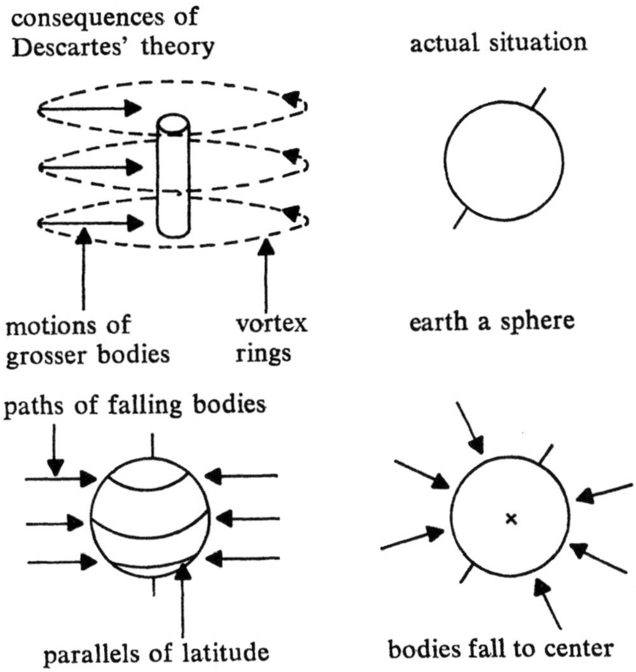


go against what is everywhere observed to happen to falling bodies. Freely falling bodies do not, observes Hooke, gravitate to parallel circles of latitude. They rather all fall toward a center point. With these two overpowering facts of experience against him, then, the Frenchman could not possibly be correct in his celestial mechanics.[32] In an effort to circumvent these difficulties inherent in the Cartesian system the Curator was led to develop his own system of celestial mechanics.

In summary let us say that it would be hard to imagine a simpler scheme of things than that postulated by Descartes and accepted by Hooke. It can be regarded as either one of the most naive or most sophisticated ventures in the history of human thought. Hooke sincerely believed that he had solved the riddle of the universe. All that remained for him to do was to now look at it carefully in a Baconian way. Soon he would know all.

Referring again to Meyerson's work may throw some light on the subject. Meyerson insists that the human mind is inexorably driven to postulate

[32] See P.W., pp. 201-202.

something persisting throughout observed changes. There must be identities in nature if nature is to be intelligible. As it happens, Cartesian mechanism is the simplest form of identity. Meyerson has tried to show that since the dawn of history mechanism and science have grown together. Mechanism is not the result of science but its *a priori* metaphysical foundation. The enquiring mind imperiously demands an ontological reality underlying and explaining observed phenomena. Mechanism, then, is not an end product of research but a means to an end, a way of satisfying our intellectual lust for unity. "It is permissible, therefore," claims Meyerson, "to state that science really tends toward the reduction of all phenomena to a universal mechanism or atomism, defining these terms so as to include electrical theories, and remembering that the causality of being, so near a relative of the causality of becoming, demands that the elementary particles be made of a single matter possessing only a minimum of qualities, in such a way that it may be, to a certain extent, identified with space or its hypostasis, ether." [33] Descartes, "perhaps the most powerful mind humanity can boast of," holds the historical position of being the man who so clearly and consistently enunciated a mechanistic philosophy that it, after being widely accepted in his own day, still holds a grip on scientific thinking today.

Hooke's insistence upon the explanatory efficacy of matter in motion is omnipresent throughout his writings. Local motions were used to explain gravity and sound.[34] He also made repeated attempts to explain all the physiological operations of men, animals and plants, as well as the nature of heat, in terms of various combinations of local motions.[35]

Let us briefly venture into optics in order to emphasize Hooke's insistence upon the explanatory power of matter in motion. Hooke's *problématique* here was to show this explanatory power operating in the case of light, both white and colored, especially in the colors observed as light passed through prisms and thin transparent plates.[36]

The three main sources for Hooke's view on light are his *Micrographia* (pp. 54ff), his criticism of Newton's 1672 paper on light which Hooke delivered to the Society on 15 February 1672, and several lectures on the subject delivered to the Society in 1682 and reprinted in the *Posthumous Works* (pp. 77ff). All the essential points, however, are given in the *Micrographia*, with the two latter sources being largely reiterations.

[33] E. Meyerson, *op. cit.*, p. 410.

[34] See Gunther, Vol. 6, 7/6/1664.

[35] See M., p. 13 and P.W., pp. xxiv, 140ff, 175-176.

[36] See for details E. Whittaker, *A History of the Theories of Aether and Electricity* (Edinburgh, 1951), Vol. I, ch. 1 and Sabra, chs. 7, 10, 11, 13.

Hooke's theory is a development of Descartes' theory. According to Descartes, light in luminous bodies, such as the sun and stars, is caused by a circular local motion of their minute internal parts. However, Descartes regarded the transmission of light away from a luminous body not as an actual local motion of the ether, but as a tendency to motion; a pressure on the minute parts of the ether affecting one point and then another in a progressive fashion. As the matter in the vortices spins, the particles closer to the center, containing some luminous body such as the sun, press outward. The pressure is passed on from particle to particle. This transmission of pressure is light. It is like someone feeling a stone with a stick. The stone does not move to his hand, but pressure is transmitted along the stick to the hand. Also, the denser the medium the faster the propagation of light. The various colors were explained by postulating different speeds of rotation of the etherial globules; the fastest give red; the slowest blue.

Hooke, on the other hand, saw no reason to deny that light was in fact an actual motion from one position to another of the minute parts of both bodies and the ether. Hooke's own view developed out of his criticism of Descartes. As he explained in his *Micrographia*, there are four kinds of local motion possible in some luminous body such as a diamond. The parts of the diamond could move in circular paths around some center. This seems to have been Descartes' position. But this would not explain how a diamond could keep its angular and pointed parts. The motion could also be that of each little part turning on its own axis. But this could not explain how the motion (light) is transmitted to the ether. Thirdly, any disarranged, irregular collection of local motions would tend in time to make the diamond fall apart and appear fluid, so this cannot be the explanation. By the process of elimination, then, Hooke is left with what he called a vibrating local motion. This is a regular motion; that is, all the parts move together in some orderly sequence of motions. Exactly what this sequence is he does not say.

His experience with diamonds told him that such vibrations must be exceedingly short and quick. Diamonds, as well as other substances such as dead fish and rotting wood, shine even though there is no actual motion visible in the form of flames. Therefore, the vibrations must be below the threshold of human sensations. Hooke is careful to add that not everything vibrates with the type of vibrations needed to produce light, hence the existence of dark objects.

The transmission of light is also by means of regular local motions. Since the whole of the universe is filled with a homogeneous (isotropic) ether without any gaps or voids, some portion of the ether is sure to be in contact with every luminous body. The vibrations in the body are therefore easily transferred to the ether. Furthermore, since the ether is so compact and homogeneous, the light is transmitted through it at an incredible, though not instantaneous, speed. Once the motions are started they spread out in right lines, radiating in all directions from the source, thus forming the shape of a sphere. He makes an analogy with the water waves started by throwing a stone into a quiet pool.[37]

Even though Hooke did not leave posterity any pictures of his regular vibrations, or give any verbal description of them, it seems rather clear, nevertheless, that he was talking about what we today would call waves. The ether is perfectly gapless and light is a regular motion of the ether. This can only be an undulating movement of different sections of the ether relative to one another. That this is what Hooke truly had in mind is given away by his analogy with water and a few words nonchalantly inserted in his "Considerations" upon Newton's theory of light. This brief report will be considered shortly. In his criticism of Newton, Hooke had occasion to mention his own view of light. As part of his explanation the Curator noted that light is propagated in the ether by simple and uniform "pulses or waves, which are at Right angles with the line of Direction." For Newton, light is a body, a stream of particles. For Descartes and Hooke, light is a motion taking place in a voidless *plenum*. For Descartes this motion is comparable to the motion of brake fluid when stopping a car. For Hooke it is a pulse propagated perpendicularly to the direction of propagation. It is therefore reasonable to think of Hooke's vibrations in terms of such a modern conception as regular transverse waves (i.e., at right angles to the line of propagation with one pulse following another at even intervals). This interpretation seems at least probable. And, if indeed it is true, it would mark Hooke as the only major wave theorist of note previous to Huygens.

Hooke also had a mechanistic explanation for the difference between white and colored light. As long as the light from a luminous body does not encounter any obstacle in the ether its motion will remain regular and propagated in a rectilinear fashion. This situation affects us physiologically as white light. But what happens when white light encounters an obstacle? If the obstacle is opaque, the light vibrations cease. If the obstacle is transparent and a light ray strikes it at an angle, the ray is bent or refracted as it passes through. This refraction occurs in a certain way. The two edges of the ray acquire two different speeds as the ray begins to penetrate the medium of refraction. That part of the vibration front which first strikes the more dense body (and all transparent bodies will be more

[37] See M., pp. 55ff. See also Sabra, pp. 185-197.

DIAGRAM ILLUSTRATING HOOKE'S THEORY OF COLORS

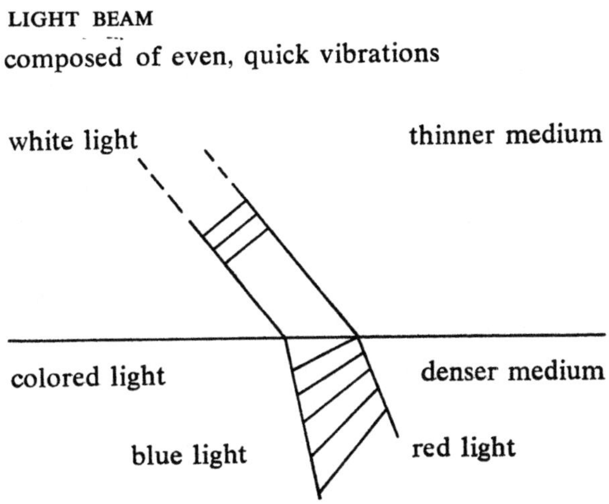


The part of the light beam that first strikes the denser medium speeds up.
The blue is therefore more spread out or "weaker."
The other colors are "mixtures" of blue and red.

dense than the ether) is speeded up by the more dense medium at the same time as the ray bends in and down (see diagram). Moreover, the part that strikes is "weakened" and "deadened" by its initial striking. This side of the ray yields blue. The other side of the ray, however, is "stronger" having struck the refracting medium after the blue edge had prepared a way for it. As the ray continues to penetrate the medium, the initial divergence in speed is also continued and the ray will spread out a little. As it does so the other colors in the spectrum appear. Hooke calls these other colors between the deepest blue and brightest red "dilutings" of blue and red. By this he meant that all intermediate colors are merely different degrees of refraction between the two extremes of blue and red. For Hooke colored light was in a sense a property of an illuminated medium in that it was the medium which split up the incoming vibration front. The colors exist only as different degrees of refraction within the medium and not as part of a heterogeneous composition which, when taken collectively, constitutes white light. This latter notion was later maintained by Newton.[38]

[38] See M., pp. 57ff. See also Sabra, pp. 254-259.

Newton's first paper on light was read before the Society on 8 February 1672. One week later Hooke's "Considerations" upon Newton's views were read. They were not published, however, until many years after Hooke's death, appearing first in Birch's history of the Royal Society. Hooke made two main points in his criticism of Newton. Hooke was willing to accept, first of all, the accuracy of the experimental results reported by Newton. Hooke himself claimed to have made hundreds of similar experiments with essentially the same results. Secondly, Hooke was not willing to admit the factuality of Newton's hypothesis explaining the results, preferring instead to remain faithful to his own theory (which we have explained above) because he thought it simpler than Newton's. Let us look at Hooke's "Considerations" more closely.[39]

Hooke understood Newton to be saying that white light is a heterogeneous aggregate of distinct and separate bodies or corpuscles (i.e., the various colors) compounded together so as to produce in us a certain sensation. When white light passes through a prism or some other kind of transparent body, its heterogeneous elements are separated thus giving us the spectrum. To Hooke's mind such a view was unduly complicated. To counteract Newton's theory Hooke presented his audience with a capsule form of his own theory (which we have explained more fully above).

He then goes on to answer an objection against his own theory which some might regard as very serious. Newton had observed that a beam of white light could be refracted into colored light by passing through one prism and could later be re-established as white light by being passed through a second lens. Does this not imply that light was originally heterogeneous in nature? Hooke did not think so. Such a phenomenon could just as well be explained, he thought, by supposing that the simple, undifferentiated local motions constituting white light according to his theory were first differentiated by refraction in the first prism (thus producing colors) only to be later re-established in their original patterns by having their differentiations cancelled out when passed through the second lens. In both explanations the same facts were used. Hooke, however, did not take Newton's claim to have "seen" the colors emerging from the convex lens merge into white light as a fact.

Hooke also showed himself to be aware of other possible theories. The Curator offered his younger contemporary two such theories in order to show that he was capable of considering views other than his own. One

[39] See Gunther, Vol. 6, 2/15/1672. See also More, ch. 4 and Sabra, pp. 251ff. In 1675 Newton delivered his second paper on light. He deliberately held up the release of his *Opticks* until Hooke died.

was to make an analogy with pigments. Perhaps it was the case that white light is produced out of colored lights as different colored paints are produced by combining other colors. That is, could it be that white is a combination of colored bodies which keep their *actual* color even after combination? Hooke could find no justification for this theory in practice, however. He said that he would be glad to hear of a case in which all the colored bodies in the world compounded together would produce a white body. He had never seen such a thing and was pretty sure he never would.

There could be another possibility, though. Perhaps white light is a combination of heterogeneous elements which lose their actual heterogeneity when compounded. In such a case colors would be *potentially* in the white light. Each color, for instance, could have its own peculiar type of local motion. But, when combined, the local motions peculiar to each could blend into some one vibrating motion or wave. That is, the pulse or wave constituting white light could be the resultant of many different vibrations, just as what appears to be the uniform motion of a body moving through the air (e.g., an arrow) is the resultant of various opposing forces. The function of a prism, then, would be to separate out and return to their actual states these peculiar local motions, thus giving us the spectrum.

In the end, however, the Curator found all theories other than his own wanting. The simplest solution he felt was to postulate a uniform wave motion for white light and variations produced in such a wave upon encountering a prism for colored lights. He was no more ready to assume colors to be already in white light, either actually or potentially, than to assume that all music is already in the air used in a pipe organ or all sounds in an unplucked string.

It is interesting to note that Hooke came close to formulating a precise wave theory of light, but never did. Nevertheless, he can and has been considered a forerunner of Huygens, even by Huygens himself. Near the beginning of his *Treatise on Light* (written about 1678, published in 1690) Huygens refers to Hooke as one of those who had begun to consider light as a wave motion. Perhaps the appraisal of Hooke's efforts with which this author can best agree, is that given by Sabra, in his thorough and sympathetic study of Hooke and his contemporaries, when he observes that because of his simplistic bent of mind Hooke seems to have always been satisfied and content with his vague, uncomplicated vibration theory. He saw no need for anything more sophisticated and so never developed his theory any further.[40]

[40] See Sabra, p. 261.

Hooke's criticism did not sit well with Newton. A series of remarks, letters and papers ensued, stimulated by Oldenburg, who cared little for Hooke personally, which culminated in a letter from Newton to Oldenburg dated 21 December 1675 in which Newton accused Hooke of doing little more than modifying Descartes' view while Newton himself did not owe anything essential to Hooke's *Micrographia.* Hooke attempted a reconciliation in the form of a pleasant letter to Newton at the beginning of the following year. However, the germ of friendlier relations was soon crushed out as a result of the controversy over gravity.

*4.2 The Properties of Falling Bodies.*

With respect to the topic of freely falling bodies close to the earth's surface, seventeenth century thinkers were faced with several main issues. One was the rate of uniformly accelerated motion. A second was momentum considered as a motive force. Another was the path that a falling body would take while descending toward a moving earth. We will here consider Hooke's views on the first two problems.

Both during and after the 1630's Galileo's work was avidly discussed on the Continent, especially in Paris and Holland. Isaac Beeckman, Marin Mersenne, Pierre Gassendi, Descartes, Giles de Roberval, Evangelista Torricelli, Pascal and Huygens were all taken up in the new natural philosophy and to some extent were disciples of the Italian.

Whatever happened on the mainland of Europe could not long be kept from England. Thomas Hobbes, as a consequence of his several trips back and forth across the Channel, carried Galileo's views to his homeland between 1634 and 1652. Before his death in 1697 Aubrey was a constant visitor in European intellectual circles. Boyle also traveled to Europe and, in 1641, visited Italy. Henry Oldenburg visited several Paris gatherings concerned with the new science, sponsored by the Dupuy brothers, between 1659 and 1660.

In 1665 Thomas Salusbury translated both of Galileo's main works into English. Hooke, undoubtedly, heard much of Galileo during his university days and by 1665, in the Preface to his *Micrographia,* could speak of him as the "famous Galileo" without in any way having to justify his adjective.

Early in his career it appears that Hooke was not totally convinced of Galileo's law for uniformly accelerated motion and, in a fashion typical of Hooke, decided to see for himself. This he did several times between 1664 and 1682. In the summer of 1664 for the first time Hooke reported to the Society that he had experimented with a lead ball from a considerable height and found that it fell fifteen and one-half feet in the first second and that the over-all rate of descent agreed well with the expected results, i.e., an arithmetic increase in time was accompanied by a geometric increase in distance fallen.[41]

But how did Hooke interpret his experimental results proving to him there was a uniform acceleration of falling bodies which in air approximated Galileo's law? His interpretation can be seen in two places, both of which reveal the same attitude. One is a debate before the Society in 1678.[42] The other is in his 1682 treatise on comets and gravity.[43] Let us note what Hooke had to say in 1678.

In the Spring of 1678 the Royal Society was discussing methods for measuring the depths of the ocean. Hooke had suggested a simple method employing a weight pulling down a container that would fill up to various levels depending upon the water pressure. Several members, however, objected that Hooke's device would not work because, based upon experiments of Galileo, the device would accelerate as it descended. On the contrary, thought Hooke, after twelve feet a terminal velocity would be reached. Hooke insisted that Galileo's law "had been made upon a theory, and not upon experiment; for that experiment would evidence the contrary. And though in a vacuity of water, air, or any other gross fluid, those proportions would hold very near; yet in a medium, wherein there was a resisting fluid body, it would not hold in any wise, especially in those, which had a considerable proportion of specific gravity to that of the descending body."[44] He recalled his experiments tried from the top of St. Paul's steeple which plainly showed a lead ball to quickly leave a wooden ball and a cork ball far behind due to the greater affects of air resistance upon the wood and cork. In order for Galileo's law to hold, one must have a very dense body in a relatively thin medium. When Hooke had this situation, as when he used a small lead ball in the air, he found Galileo's proportion to hold pretty near. In addition, claimed Hooke, as the object's speed increases so does the resistance of the medium as could be observed in the cases of birds flying and oars breaking by too swiftly striking the water. Furthermore, regardless how thin the medium was, sooner or later a terminal velocity, differing with different bodies, would be reached. He does not mention that Galileo also recognized this fact.

[41] See Gunther, Vol. 6, 8/17/1664.
[42] See Gunther, Vol. 7, 4/4/1678.
[43] See P.W., p. 182.
[44] Gunther, Vol. 7, 4/4/1678.

After this "the progress of the body would always be made by equal spaces in equal times, though ever so far continued, provided the gravitating powers remained the same."[45]

With respect to momentum considered as a motive force, the Curator once again shows himself to be the practical-minded scientist. Because it would be useful, Hooke set out to investigate the relationship between the motive force and velocity of bodies. As Hooke expresses himself, experiments of this kind would be very useful in mechanics, "so could they be made with bodies perfectly solid, would they be for the establishment of one of the chiefest philosophical principles, namely, to show the strength, which a corpuscle moved has to move another."[46] This is Hooke's original problem. To solve his problem Hooke designed several experiments. These are preserved for us in the records of Hooke's work before the Royal Society for 1663 and 1669. Since there is some development in Hooke's thinking, we must review both years' work.

For his 1663 experiment, Hooke's apparatus consisted of a pan balance and a set of small weights. In the first part of the experiment, the Curator dropped a one ounce weight onto one pan of the balance while the other pan contained in succession various weights. He found that from .191 inches the falling one ounce weight could move a maximum weight of four ounces. A fall of .667 inches could move a weight of eight ounces. Hooke continued his trials using counterweights of 16, 32, 48, 64, 96, and 128 ounces. He then repeated the whole procedure employing a falling one-quarter ounce weight. Hooke found, by way of a general conclusion, that he had to quadruple the height of fall, in order to double the velocity, in order to move twice as much weight in the balance pan. In other words, if a weight $w$ falling a distance $d$ could move a weight $W$, then the same weight $w$ falling $4d$ (thus moving twice as fast according to Hooke) could move $2W$. Although he realized the shortcomings of his apparatus Hooke nevertheless felt safe in concluding that the trials "though they do not answer our expectations as to the accurate exhibiting of the strength of a moved body, yet seem to prove, that a body moved with twice the celerity acquires twice the strength, and is able to move a body as big again," that is, twice as heavy.

In the beginning of 1669, Hooke again attempted to demonstrate the same thesis using basically the same method as the one employed in 1663.

[45] *Loc. cit.* In fairness to Galileo it should be pointed out that he also recognized the existence of terminal velocities for all falling bodies. In the *Dialogues Concerning Two New Sciences* (First Day, Section 119) Galileo admits that his law is an ideal case which would not hold for a body descending in a medium.

[46] See Gunther, Vol. 6, 2/18/1663.

This time, however, it was tried out of doors with greater weights and heights. The first 1669 attempt failed, claimed Hooke, due to frost on his instruments.

The next step is to formulate Hooke's work mathematically, something he himself did not do. Galileo had said that the motive power or momentum of a body was directly proportional to its speed at the time of impact ($V = M$). Hooke also had heard from the Italian that a falling body (which in practice would have to be very dense relative to the medium) would have to fall four times as far in order to double its velocity. In the *Dialogues Concerning Two New Sciences* (1638), The Third Day, Galileo defends the position that a body falling from rest traverses distances in the same ratio as the odd numbers beginning with unity during equal time periods. In other words, the falling body will fall faster and faster as it descends according to a definite geometric ratio (1, 3, 5, 7, etc.). This means that the distance covered in each succeeding equal unit of time will increase at a much greater rate than the total amount of time elapsed. This relatively greater increase in distance also occurs at a definite rate, namely, it is directly proportional to the elapsed time squared. The velocity of the falling body, therefore, does not increase in proportion to the space covered as a naive person is likely to believe. That is to say, someone might think that the velocity (and also the motive power) of a falling body will double when the distance fallen is doubled, triple when the distance is tripled, etc. However, this is not the case. It is rather the case that one must quadruple the distance fallen in order to double the velocity. For example, at the end of a 16 foot fall the body's speed will be 32 feet per second, at the end of a 64 foot fall it will be 64 feet per second, at the end of a 256 foot fall it will be 128 feet per second, and so on.

This is the background against which Hooke seems to have been working. Although he did not make any theoretical advances beyond Galileo, his 1663 experiments tended to confirm Galileo's view on the distance covered – velocity relationship as well as that between velocity and motive power. These two factors can be combined into the relation $NV_0 \longleftrightarrow N^2D_0$. This formula summarizes the following correspondence: to an original velocity $V_0$ let there correspond an initial distance $D_0$; then, to a velocity increased N times, namely $NV_0$ there is a corresponding distance $N^2D_0$. Therefore, to obtain twice the velocity requires four times the distance, three times the velocity requires nine times the distance, etc.

In order to make what Hooke had in mind crystal clear, let us assume that one wants to double the speed that a body has after falling four feet. The result would be sixteen feet. This means that the body's velocity and momentum will double when its original distance of four feet has been quadrupled (i.e., raised to sixteen feet). If one wishes to triple the velocity, assuming one foot as the original distance, one would have to increase the height to nine feet, as shown in the formula $3V_0 \longleftrightarrow 3^2(1)$.

Hooke also attempted to bring into play the relationship between speed and weight. Galileo's formulations had ignored weight. It must be recalled that Hooke realized that Galileo's views were set in an ideal situation. In a vacuum perhaps one could ignore the weights of falling bodies. In real life, however, the relative weights of bodies are important. Hooke knew from his own experiments (not the ones now under discussion) that a lead ball would soon leave a cork ball far behind as they fell.

Consequently, in his second 1669 attempts, Hooke changed his tactics and attempted to show that a quadruple weight would be required to double the velocity of a body if the distance fallen is constant. Here he is concentrating on the relationship between weight and velocity rather than the relationship between height of fall and velocity. In each case, though, the velocity would double. To show this he experimented with a pendulum and with running water. He found that with the time and distance of swing constant pendulum weights of 2, 8, and 32 ounces would produce 12, 24, and 48 vibrations respectively. He also found that the volume of water in his vessel had to be four times as large in order to run out of the container twice as fast. This last experiment was not too successful, noted Hooke, because his vessel leaked.[47] Although he did not formulate his work mathematically or, as far as is recorded, attempt to perfect his experimentations, Hooke was convinced that if a weight $w$ moving with velocity $v$ could move a weight $W$, then $4w$ would move with $2v$ thus being able to budge $2W$. To express these later experiments mathematically, one would simply replace the $D_0$ in the above formula with $W_0$. The $W_0$ would then represent the original weight. Thus: $NV_0 \longleftrightarrow N^2W_0$. Again the N stands for the number of times the velocity is to be increased.

Again, the force of momentum is directly proportional to the velocity (not the velocity squared). The faster moving pendulum or water would strike twice as hard as it would originally. Now, however, Hooke knows how to double the velocity by geometrically increasing the weight.[48]

[47] See Gunther, Vol. 6, 1/14/1669. In *Lampas* (1677) Hooke gives accounts of more successful water trials. See Gunther, Vol. 8, pp. 181ff.

[48] L. D. Patterson's attempt to make Hooke (along with Leibniz) into a precursor of the formula of kinetic energy (K.E. = $wv^2/2g$) is forced. Hooke refers explicitly to distance and weight, not energy. Also, for Hooke M or F is proportional to V and not $V^2$. See her "Robert Hooke and the Conservation of Energy," *Isis*, Vol. 38 (1948), pp. 151-156.

The Curator later attempted to expand upon Galileo's thought as well as upon his own 1663 and 1669 work. He maintained that in many other areas the same relationship will hold between the factors related; namely, a geometric increase in one is a function of an arithmetic increase in the other. In some remarks added at the end of his *Lampas,* first published in 1677 and later made a part of his *Cutler Lectures,* the Curator lays it down as a fact that the actions of bullets, arrows, slings, pendulums, musical strings, vibrating bodies and falling bodies adhere to his formulation. In Hooke's own words, "if any Body whatsoever be moved with one degree of Velocity, by a determinate quantity of strength, that body will require four times that strength to be moved twice as fast, and nine times the strength to be moved thrice as fast, and sixteen times the strength to be moved four times as fast, and so forwards." Hooke's view can be seen to conform to the formula stated above: $NV_0 \longleftrightarrow N^2(D_0 \text{ or } W_0)$. In his 1663 trials, for example, multiplying the height fallen (which can be considered a "quantity of strength") by four increased the velocity by two, while in his later 1669 experiments increasing a pendulum's "quantity of strength" from a 2 ounce weight to a 32 ounce weight (16 times) increased the number of vibrations from 12 to 48 (4 times). To Hooke's way of thinking the "quantity of strength" (e.g., D. in 1663; W. in 1669) decided a body's velocity at the time of impact. The velocity, in turn, decided the force or momentum of impact. By continually increasing the velocity one could move continually heavier bodies, a useful thing to know. Hooke regarded his having verified this as the prime importance of his work.

Hooke's work on impact in 1669 took the Curator off on to a side track which ultimately resulted in a dead end. There is no accepted modern formula expressing what Hooke did. However, even though he did not state anything new and positive, he did manage to contradict Descartes' fourth law of impact stated in the *Principles of Philosophy,* book II, section 49. As can be seen from the context of both years' work, Hooke was conscious of this denial, even though he did not deny Descartes' more general third law of nature. Also, this appears to be the only one of Descartes' seven laws of impact which Hooke publicly discussed. According to Descartes' fourth law of impact, which he claimed to be a deduction from his third law of nature, if C is completely at rest and is larger than B, then B will never move C regardless of how fast B hits C. Merely being larger in size gives C greater resistance to being moved than B has force to move it. Since Descartes' laws are not valid, Hooke's denial of this law can be considered a minor contribution to mechanics.

But did Hooke's work in any way influence those who came after him?

It is possible to see Galileo's and Descartes' well-known momentum formula, i.e., the force of impact of a body is a function of its weight or size (not mass) and velocity (F = wv), emerge from Hooke's experiments. Although he does not have weight as a factor in his V = M formula, it appears to be there as a constant in any particular case. Let us take, for example, a body weighing ten pounds moving at ten miles per hour (M = wv; M = (10)(10/1) ). Its motive power would be 100 units. Now imagine the same body with double the velocity (M = (10)(20/1) ). Its motive power would now also be doubled, being 200 units. The same would be true of any other constant weight. This appears to be about as far as the Curator carried his investigations.

The question now remains: Did Hooke in any way influence Newton's second definition (P = mv) or second law of motion (F = ma) stated at the beginning of his *Principia?* As part of his explanation of Law II, Newton states: "If any force generates a motion, a double force will generate double the motion, a triple force triple the motion, whether that force be impressed altogether and at once, or gradually and successively."

In John Herivel's *The Background to Newton's Principia*, we have documented the close correspondence between the manner in which Descartes expressed himself on certain basic notions in mechanics and the way in which Newton, in his early manuscripts (starting about the end of 1664), expressed himself on the same subjects. Based upon this evidence, Herivel concludes that there was a direct influence of Descartes on Newton with respect to inertia (Newton's first law of motion) and momentum, and an indirect influence with respect to the concept of force. The derivation of force (Newton's second law, F = ma) is indirect because it involved an important modification by Newton of Descartes' (and Galileo's) momentum equation. Where the concept of momentum (i.e., wv) came from originally (perhaps Buridan) is unknown.[49]

Now let us recall the time sequence between Hooke's first experimental work on velocity and motive power and Newton's earliest manuscripts. Hooke's first recorded work, in which he states (at least implicity) the formulas given above, occurred in February of 1663. Newton's earliest manuscripts, consisting of notebooks or "waste books," in which the topics of impact and momentum first arise, date from the end of 1664. Could Newton have obtained the first hint of these notions from Hooke? True, there is no mention of Hooke in Newton's earliest known notebook (listed as MSI by Herivel) while there is a reference to Descartes' *Principles of Philosophy*. Yet, Hooke's work was publicly presented to the Royal Society. And,

[49] See Herivel, ch. 2.

there was a reasonably sufficient time lapse during which Newton could have been informed of it.

Although Hooke said no more than Descartes concerning momentum, he did, in opposition to Descartes' *a priori* method, give some experimental verification which would have impressed Newton, perhaps stimulating him to accept as true what later became his second definition. Also, there seems to be some hint of Hooke's direct correspondence rule ($V = M$) contained in part of Newton's explanation of Law II. Hooke had said in 1663 that if a body with a certain velocity (what we may call a force) can move another body with a certain weight (what we may call a motion), then twice the "force" will generate twice the "motion." Newton, in the first part of Law II, states that "The change of motion is proportional to the motive force impressed." Hence, if a force can produce change in a motion, double that force can produce double that change in motion (i.e., $a = F/m$ so that when m is constant, a and F are directly related). The correspondence, of course, between Hooke and Newton is not exact. Newton is speaking about what it takes to produce in a body a change in motion (i.e., an acceleration) while Hooke is asking how much must a body's constant velocity at the time of impact be increased so as to double its ability to move another, heavier, body.

Perhaps one way of bringing Hooke and Newton closer together is to regard Hooke's experimental work and Newton's second law as two different approaches to the same phenomenon. While Newton's approach is from the point of view of the effect, the body which undergoes a change in velocity, Hooke's approach is from the point of view of the cause, the body which produces an acceleration in another body by being itself moved at double its original velocity.

At the present time, there is no conlusive evidence that Newton was influenced by Hooke in conjunction with Descartes and Galileo. However, the *possibility* of an influence of Hooke on Newton in this matter appears well founded.

### *4.3 The Springiness of Matter.*

Finding a perfectly hard or solid body was impossible, thought Hooke. This was because all bodies contained relatively less dense parts into which the relatively more dense parts of the body could be compressed. Even the hardest substances, such as glass, contain air. Therefore, all bodies can be compressed. Moreover, he claimed, a perfectly hard body would not bounce at all. To test this, one could drop progressively harder bodies to see if there was a steady progression in lack of bounciness. It does not appear, though, that Hooke ever actually carried out his plans.[50] But he did perform two different experiments along the same lines.

In one experiment, to show that wood is not perfectly solid, he hung three wooden balls, by equal lengths of string, a few inches apart from one another, in the same vertical plane. He found that when he raised one end ball and let it fall against the middle ball, the middle ball would not move very much but the other end ball would fly up a distance as far as the originally moved ball had been raised. Hooke's only conclusion was that wood is not perfectly solid. Some members of the Society, however, thought it might be used to support Hooke's belief that the total amount of motion in the universe was constant; that is, the same now as at the time of creation. Some things start to move while others stop, but the total is always the same.[51]

Two weeks later, Hooke impelled wooden balls against springy and less-springy bodies in order to show the Society that the degree of bounce of the wooden body was proportional to the degree of springiness of the hit body. However, since this was not a precise way of measuring the springiness of the solid bodies used, neither the Society nor its Curator was satisfied with the results. It was on this occasion that Hooke proposed experimenting with a metal spring, over which one could have much more precise control, in order to test the relationship between degrees of weight and tension.[52] This mode of experimentation was later used to derive Hooke's Law.

It now remains for us to consider the culmination of Hooke's terrestrial mechanics. In his drive to understand nature as it really is, the Curator was led by Bacon to experimentation and by Descartes to matter in motion. Matter, motion and ether were not problems for Hooke. Rather, they were

[50] See Gunther, Vol. 6, 10/29/1668. Hooke does not seem to have been directly interested in or to have worked upon the problem of impact as such. Other members of the Society did take up the problem. Wallis' brief report to the Society in November of 1668 concluded that two inelastic bodies of the same weight moving in the same direction with the same speed would merge into one and continue with the same speed. If they were moving in opposite directions they would merge and come to rest. Wren reported to the Society in December that two perfectly elastic bodies, coming together with equal velocities on the same line, would transfer their velocities and directions. See *Philosophical Transactions*, Vol. 3 (1668), pp. 864ff. Interest in the subject was originally stimulated around 1600 by a desire to know how a falling body would strike a moving earth. Other work on the subject was done by Galileo, Descartes, Borelli, Gregory, Huygens, Mariotte, and others. Newton summarized the results in his *Principia*, bk. I, Scholium to the Axioms, or Laws of Motion.

[51] See Gunther, Vol. 6, 11/12/1668.

[52] See Gunther, Vol. 6, 11/26/1668.

the principles by which to solve problems. We have seen what some of these problems were in pneumatics and mechanics. But, Hooke asked himself, was there some way of unifying his previous work under some general law? He thought there was.

Hooke's Law was first published in 1676 in the form of a mysterious collection of letters appended to Hooke's *A Description of Helioscopes.* The letters were *ceiiinosssttuu.* Two years later Hooke published his *De Potentia Restitutiva, or of Spring Explaining the Power of Springing Bodies.* This latter was then published a year later as the last work in a group of works composing Hooke's *Cutlerian Lectures.* In this work on springiness the letters revealed in 1676 were rearranged so as to read *ut tensio sic vis;* that is, as the extension so the force, or, in more modern terminology, strain is proportional to stress.

The immediate source of Hooke's Law was a series of experiments performed by Hooke upon, principally, a helical spring. The remote sources of his law, however, as can be seen from the applications of the law, are many and varied. His general law appears to have been a unifying principle epitomizing in a neat, succinct way all previous work, both by himself and by others, on the relationship between two independent variables which bear a mutual cause and effect relationship.[53] Hooke states his conclusion for us: "The Power of any Spring is in the same proportion with the Tension thereof: That is, if one power stretch or bend it one space, two will bend it two, and three will bend it three, and so forward." [54]

As Hooke states further, the experimental work behind his law is very simple. It consisted of two simple procedures. One procedure was to take a wire, coil it around a cylinder to the form of a helix, and then suspend it by one end from a hook. On the other end Hooke hung various weights, noting the amount of stretch undergone by the spring in each case. He observed that there was a direct proportion between the weight suspended and the increase in the spring's length. "And this," Hooke confidently declared, "is the Rule or Law of Nature, upon which all manner of Restituent or Springing motion doth proceed."

Another procedure was to suspend a twenty to forty foot straight wire by one end and then proceed to attach weights to the other end. As the line stretched under successively heavier weights, one could notice that

[53] Galileo, in his *Dialogues Concerning Two New Sciences,* Second Day, had dealt with the relationship between forces and breaking points of variously shaped bodies. Hooke also seems to have had some interest in the strength of materials. See Gunther, Vol. 6, 7/15/1669.

[54] Gunther, Vol. 8, pp. 333-334.

the several stretchings of the said wire "always bear the same proportions one to the other that the weights do that made them." [55]

Hooke foresaw innumerable applications for his principle. The making of weapons, for instance, such as bows and catapults, would be greatly facilitated. His earlier work on the compression of air was declared to be a special case of his newly discovered principle. In addition, he could see his V = M formula as an example of this principle. Also, we find an isolated assertion by Hooke stating that the resistence of fluids to bodies moving through them decreased or increased in a continual proportion as the body decreased or increased its velocity.[56] For Hooke, then, in any interaction involving a springy body, one degree of mechanical pressure (a push or pull) would produce one degree of effect; two degrees of pressure, two degrees of effect, etc. The kind of effect produced and its extent would, of course, vary with the material involved.[57]

All of these applications, though, important as they were, were not the most important application of his work, because, to Hooke's way of thinking, they were not the most practical. As late as 1678, as we see here, Hooke was still interested in explaining the construction of a dependable clock that could be used at sea. As a further step in Hooke's Law, the Curator hoped to reverse the process of stretching a spring. That is, he thought he could deduce from his work the possibility of compressing a spring in a watch and having it unwind at a constant rate thus providing the basis for a workable clock unaffected by storms, humidity, changes in gravitational pull, etc. This he thought he had accomplished. If the extension of the spring is directly proportional to the distorting force applied ($S \propto s$), he thought it to follow that the restoring force in a compressed spring would also bear a directly proportional relationship to the spring's decompression. This means that the decompressing spring will move equal spaces in equal times. It is this isochronism, of course, which so attracted Hooke's attention. [58]

[55] Gunther, Vol. 8, p. 335.

[56] See Gunther, Vol. 6, 12/31/1668.

[57] See Gunther, Vol. 8, pp. 336ff.

[58] We do not wish to become involved in the debate, already mentioned at the beginning of our study, over who first invented the spring-driven pocket watch. The argument between supporters of Huygens and Hooke has been going on for three hundred years, is fairly well exhausted, and we can add nothing to it. We know that Hooke claimed to have such a clock as early as 1658, that both showed working models about 1674, and that Hooke's Law, containing the basic principle used in such watches, was first published in 1676. Although none of the watches produced in his lifetime were very good, it is probable that Hooke was more responsible for the basic idea behind such a watch than anyone else. But who actually was the first to produce

Hooke's tension law was not definitive, however, just as "Boyle's Law" was not. The Curator had failed to distinguish between elasticity and plasticity. As far as Hooke knew, all bodies would respond to pressure according to his principle until they reached their breaking point. Hooke failed to realize that it was possible to bend an object and have it stay bent instead of springing back to its original shape. Also, he did not mention the breaking point as a limit to his principle. Perhaps such considerations were so obvious that Hooke did not feel any need to mention them specifically. In any event, later thinkers found it necessary to modify Hooke's Law by specifying the exact limits within which his direct proportion law held for each particular kind of substance. As E. Williams states at the very beginning of his article on Hooke's Law, "The first law of elasticity for solid bodies, Hooke's *ut tensio sic vis*, stood as a one-sided and incomplete statement of objective truth. Its completion and full determinateness was later made by placing the so-called and hardly less important elastic limit upon it."[59]

a working model may never be known. Cf. Peter G. Tait, "Hooke's Anticipation of the Kinetic Theory and Synchronism," *Proceedings of the Royal Society of Edinburgh*, Vol. 13 (Nov. 1884–July 1886), p. 118: "While collecting materials for a Text-book of the *Properties of Matter*, the author had occasion to consult the very curious pamphlet by Robert Hooke, entitled *Lectures de Potentia Restitutiva, or of Spring* (London, 1678). In this work there is a clear statement of the principle of Synchronism, which was applied by Stokes to the explanation of the basis of Spectrum Analysis. There is also a very remarkable statement of the elmentary principles of the modern Kinetic Theory of Gases, the first mention of which is usually fixed sixty years later, and ascribed to D. Bernoulli in his *Hydrodynamica* (Argentorati, 1738)." Others of note have not appreciated Hooke's contributions to kinetic theory. See, for example, E. J. Dijksterhuis, *The Mechanization of the World Picture* (tr. by C. Dikshoorn, Oxford, England, 1961), p. 457.

[59] See E. Williams, "Hooke's Law and the Concept of the Elastic Limit," *Annals of Science*, Vol. 12 (1956), pp. 74-83 for a study of Hooke's faults and the later development of Hooke's Law.

CHAPTER V

# THE MECHANICS OF CELESTIAL LOCAL MOTIONS

## *5.1 The Cause of Planetary Motions.*

Hooke believed that the curved course of a planet about the sun was the resultant of a combination of two straight-line motions. Where Hooke originally acquired the idea that a planet's path is the combination of two motions is not clear. Newton once suggested in a letter to Halley (20 June 1686) that Hooke very well may have acquired it from Borelli. Newton's suggestion, however, has been exploded by Angus Armitage and Alexander Koyré.[1]

Giovanni Borelli (1608-1679), a professor of mathematics at Pisa, published his *Theoricae Mediceorum Planetarum ex causis physicis deductae* at Florentiae in 1666. The *imprimatur* is dated 26 February 1666. The purported purpose of the work was to explain the paths of Jupiter's satellites around their primary. These satellites were given the name Medicean Planets by Galileo, their discoverer, in honor of his patrons, the d'Medici family. It is clear, however, that he also believed that the planets revolved about the sun.

According to Armitage and Koyré, Borelli's doctrine on planetary motion was a combination of the views of Kepler and Galileo.[2] From Kepler he learned of the sun's power to move bodies; from Galileo he learned that a planet once put into motion will continue moving indefinitely (in a circle) even when the original force is removed and no new force is applied to it. It was Galileo's conjecture that God originally fashioned all the planets in some remote region of space. He then propelled them toward our sun. When each planet had reached its preordained speed and distance about the sun, God changed their rectilinear motions to their present curved paths. For Galileo, then, that the planets should continue about the

[1] See A. Armitage, "Borell's Hypothesis and the Rise of Celestial Mechanics," *Annals of Science,* Vol. 6 (1948-1950), pp. 268-282 and A. Koyré, *La Révolution Astronomique* (Paris, 1961), pp. 461-520. Herivel (p. 59 n. 4) dismisses Borelli's contributions to astronomy in one brief footnote.

[2] See Armitage, *art. cit.,* pp. 269-271 and Koyré, *op. cit.,* p. 466.

sun in the way they do demanded no special explanation. Galileo was still enough in the Greek tradition to regard circular motion as "appropriate" for celestial bodies. As we will see, Borelli did not entirely agree with Galileo.

Borelli offered three possible solutions to planetary motion, the first two of which he rejected. First, there might be some actual physical bond between the planets and the sun. But there is no physical evidence of this. Secondly, the planets might be floating in an ether that was denser near the sun so that each planet floated in its own layer depending upon the planet's density. (For Hooke the ether was thinner toward the sun.) This, however, was also hard to understand from the physical point of view.

Borelli's third alternative was to suppose that the planets were maintained in their orbits by three forces acting simultaneously. First of all, each planet has a "natural appetite or instinct" to approach the sun in a straight line. We see here a touch of anthropomorphism: the planets have something akin to desire or will power. In contrast to the Italian, Hooke would have nothing to do with anthropomorphic explanations, even in part.

Secondly, he supposed that sunlight was corporeal and capable of exerting force on the planets so that, as the sun rotated, it would physically push the planets around. Unlike Kepler, Borelli did not demand a constant application of sun-force but, instead, allowed for the possibility of a cumulative force so that each impulse remained impressed upon the body. It is here that he agreed in part with Galileo. For both men, once a body is put into motion it will continue moving even after the original force has been removed and no new force is applied. Any new force will then increase its motion. By this method Borelli hoped to explain how the seemingly feeble sunlight could in time push around heavy planets.

In the third place Borelli supposed an outward impulse, caused by the planet's revolution around the sun, which tended to impel the body *directly* away from the sun. Borelli used the famous analogy of a stone being whirled around in a sling. In this particular Borelli believed that a body would fly off in a right line directly away from center *if the cord were to be broken.*

It might well be queried why, after accepting Galileo's view on the conservation of circular motion in the heavens, Borelli needed any other explanations or forces to account for planetary motions. The answer is that by Borelli's day anyone attempting to explain heavenly motions had to take into account at least one of Kepler's findings, namely, that the planetary paths are ellipses, not circles. In the early part of the seventeenth century one could be excused for accepting circular paths. By 1666 this was no longer possible. Borelli, therefore, had to bring in other factors in order to force the planets out of their previously, very comfortable, circular orbits.

For Borelli a planet is kept in its orbit because the first and third factors are balanced for each planet. He attempted to explain why each particular planet is sometimes closer and sometimes farther away from the sun as it travels around by postulating a fluctuating motion (closer to and farther away from the sun) on the part of each planet due to the two forces fighting one another for supremacy. Also, Borelli did maintain that although the tendency of each planet to approach the sun was the same for each planet, the repulsive power did vary inversely as the distance (not as the square of the distance).

To claim that Hooke learned anything from the mathematician of Pisa would be rash. First of all, Hooke's paper on curved motion and Borelli's book were both published about the same time. Secondly, Hooke's doctrine differed from Borelli's in important ways as we will soon observe. As both Armitage and Koyré agree, Borelli's solution to the problem of planetary motions can in no way be interpreted as supplying the basis for Hooke's explanation. How Newton could think that it did remains a mystery to this day.[3]

In passing on to the Curator's explanation, we also pass into a view of the universe quite different from Borelli's. Hooke had to arrange his explanation around two basic doctrines. One was his own view on the force of gravity. The other was the Cartesian principle of inertia. Both of these basic notions were lacking in Borelli's thought.

Hooke abhorred animistic or anthropomorphic explanations in natural philosophy. This was especially true in his doctrine on the cause of gravity. As we will shortly see in detail, Hooke's view on gravity was totally me-

[3] See Armitage, *art. cit.*, pp. 281-282 and Koyré, *op. cit.*, p. 512 n. 27. If Hooke were to be accused of purloining from anyone, it might be Jeremiah Horrox or Horrocks (1619-1641), a minor English astronomer. In a letter to a friend (25 July 1638) Horrox used a circular pendulum to illustrate the motion of a planet around the sun. If maneuvered properly, the pendulum could be made to describe an ellipse. The sun, he claimed, was both driving the planets around and drawing them to itself as it rotated. Horrox thought that the two factors working against each other, plus the sun's tendency to repulse more than it attracted, would produce an elliptical orbit. It would appear that Horrox did not possess the principle of inertia either. Horrox's works were published in 1673 by John Wallis but in the interim between Horrox's death and 1673 his manuscripts were widely dispersed throughout England. If Hooke did learn anything from Horrox, or anyone else for that matter, he did not mention it. See J. Horrox, *Opera Posthuma* (London, 1673), pp. 312ff.

chanistic in character. With respect to the motions of planets, Hooke believed that the planets are gravitating toward the sun and would soon fall into it if something did not prevent them from so doing. What is this intervening factor?

To Hooke's way of thinking he could not call upon spirits, intelligences, or the like to hold the planets away from the sun since these were ruled out by his mechanistic philosophy. Nor did he find anything natural or appropriate about curved or circular motion. He did, however, following in the footsteps of Descartes, believe that there was something natural and God-given about rectilinear motion. That is to say, he accepted the principle of inertia as applying to the heavenly bodies. It was by putting together the force of gravity with the "force" of inertia that Hooke explained the planetary paths. His thinking on the subject came to a head in 1666.

In the Spring of 1666, Hooke read a very short paper on motion in a curve to the Society entitled "Concerning the Inflexion of a Direct Motion into a Curve by a supervening Attractive Principle." Why do the planets continue to move around the sun rather than moving away into space questioned the Curator. According to the principle of inertia, such a thing should not happen. Also, why do the planets continue to move around the sun rather than falling into it? According to his view on gravitation, such should not occur either.

Two possible theories presented themselves to Hooke. One was that the ether filling the area around the sun and planets was less dense near the sun, due to the expansive power of the sun's heat, and more dense as one proceeded farther away from the sun, thus causing the planets to be pushed toward the sun by the heavier outer layers of the ether. Secondly, there could be an attractive power toward the sun counteracting the straight-line motion of the planets (or *vice versa* depending upon how one wants to look at it). Either alternative, thought Hooke, could explain the phenomena in a neatly mechanistic fashion. The second alternative, however, was considered by Hooke as the more likely, although he did not say exactly why at the time. We can see, though, that the first view would deny the isotropic nature of the ether needed by Hooke to explain light.

To illustrate by analogy how the two factors could balance out to a curved path, Hooke used a conical or circular pendulum. This consisted of a large wooden ball hung by a wire from the ceiling and made to rotate in a circular path which cut a plane parallel to the floor. Hooke declared that according to his model it was possible to illustrate in a graphic way, as nearly as one could, given the crudity of the mechanical device, how it was that the planets circumvented the sun. Hooke's experiments showed that if the ball was pushed in such a way that its tendency to move away from the pusher was stronger than its tendency to move toward the center of its swing (where it would hang straight down), then an ellipse would be generated whose longer diameter would extend away from the pusher. If, however, the initial push was weaker than the weight's tendency to the center, the ellipse would have its longer diameter stretched out across the front of the pusher. If again, both tendencies were of equal strength, the path of the ball would be a circle.

In an extended version of the same type of experiment, the Curator fastened a small pendulum to the bottom of the large wooden ball in order to illustrate the moon going around the earth while the earth went around the sun. He found that it worked out pretty well but not as perfectly as he would have liked. The model and its motions appeared to approximate fairly well what was thought to actually happen in the heavens but there were also several odd little motions which Hooke attributed to unavoidable experimental error and, consequently, dismissed as insignificant.

It would seem that Hooke did not fully realize that his simple experiment had two main shortcomings, either of which would have disqualified it as a good analogy with planetary motions. One is the fact that in the case of the conical pendulum the farther out from perpendicular the bob is swinging the greater is its tendency to return to center. Now, as we know today, and as Hooke knew in a vague way in 1666, the force of gravitational attraction diminishes as a heavenly body is removed farther and farther from the central body. His pendulum experiment, therefore, would seem to work against the law of universal gravitation rather than to support it.

Secondly, the planets are known to circumvent the sun in slightly elliptical paths with the sun located in one of the foci. Hooke found that by pushing the bob in various ways he could get it to describe various kinds of elliptical paths. There is no indication, however, that Hooke was able to maneuver his pendulum into an elliptical path with one of the foci directly below the point of suspension. In Hooke's experiment the perpendicular through the point of suspension would cut through the center of the ellipse rather than through one of the foci.

All one can really say about the significance of Hooke's experiment is that at this point in his career the Curator certainly realized in a vague way that the planetary paths were a combination of motions. His advance over previous thinkers resides in the fact that he correctly identified the specific types of motion involved. On the one hand there was inertial motion while on the other there was a tendency toward center. Somehow these two motions balance out to an elliptical path around the sun.[4]

Many years later Hooke referred back to his earlier work on curved motion. On 3 June 1685 he read a discourse to the Society in which he attempted to explain the appearance of mysterious lights in the heavens to the north which usually occurred during the winter months. The phenomenon, thought Hooke, could be explained by assuming that part of the earth's atmosphere was left behind in the ether during perihelion to be picked up again during aphelion. The ether, supposed the Curator, is quiescent and so cannot carry the planets around the sun. What does keep them in their orbits is an "imprest direct Motion, and an attractive or protruding impulse towards the Center of the Sun." [5] Hooke regarded his view as original with himself and there appears to be no reason for denying that he was, subjectively at least, honest in his claim.

## *5.2 The Cause of Gravitation.*

During the course of his career, Hooke entertained two explanations of the cause of gravity: the magnetic and vibratory. The former was inherited from Gilbert and Bacon. The latter was of Hooke's own creation. The topic of gravity, and especially the question of the physical cause of gravity, seems to have been of major importance to Hooke since the beginning of his life with the Royal Society. As he stated in one of his reports to the Society, although gravity is one of the most universal phenomena in the world, it has not been given the attention it deserves. Its cause has, in fact, been neglected for centuries. It is only now (in Hooke's own time) that the

[4] See Gunther, Vol. 6, 5/23/1666. Huygens' law of centrifugal force ($F = mv^2/r$) was first published in 1673. It is possible that Huygens, hearing of Hooke's reports, was stimulated to publish his basic notions at the end of his work on clocks.

[5] P.W., pp. 196-197. As far as the ultimate cause of planetary motion is concerned, Hooke was convinced that one need go no further back than God. It was God who originally set up the motions of the planets by combining the component motions into closed elliptical paths. It is the duty of the philosopher of nature to discover what the Lord of nature has done. It appears, furthermore, that Hooke had no idea of the vast length of time the universe had been in existence. Hooke said, in a discourse he read on carriages, that he did not know who invented the wheel. As far as he could tell it was first mentioned in the Bible when Joseph asked to ride in Pharaoh's chariot. He also wondered why the "Americans" did not know of the wheel before 1492 if all people originally came from the Garden of Eden. In any event, it was invented a long time ago when the world was young; perhaps thousands of years ago. It must also have been thousands of years ago, he thought, that God set the planets in their courses. See Gunther, Vol. 7, 2/25/1685.

situation has improved somewhat. Gilbert and Bacon, for instance, discussed gravity as a form of magnetism, while Kepler made it an inherent property of all celestial bodies.[6] Hooke was not happy with these views because they contradicted experience in one way or another. Nevertheless, he was very much interested in solving the problem of gravitation. For Hooke, this meant determining its physical cause.

At first, Hooke thought that the earth was indeed a large magnet as were, perhaps, the other centers of gravity in the universe. In an effort to establish this view, he performed a series of experiments over a period of twelve years. He finally abandoned it in favor of another theory. This other theory was the vibratory theory which he was in the process of developing even while experimenting with the magnetical theory. In the end, Hooke came to explain even magnetism in terms of vibrations.

The sources for Hooke's views on the magnetic theory of gravitation are to be found among the records of the Royal Society, beginning at the end of 1662 and going through to the Spring of 1674. In these reports we see Hooke casting around in one direction and then in another in order to test the magnetic theory. Since each attempt is a different type, although directed toward the same end, we will review each separately in chronological order. In addition, we should mention a passage in the *Micrographia* which throws light upon Hooke's early magnetic view of gravitation.

At the end of 1662, the newly appointed Curator set out to test the effects of gravity on a body raised above the earth. Gunther records the results for us. According to the account given by Hooke to the Society on the decrease of gravitational pull as a body was raised above the earth's surface, Hooke was curious to see if such a decrease really occurred. Assuming that the earth was a great magnet, it would follow that the pull of gravity would decrease as one went farther *down* into the earth, because the magnetical attraction of the above parts of the earth would counteract the effects of the lower parts. Hooke reasoned that this same decrease in weight should occur if the body was taken farther away from the magnet altogether. It was clear from operating with small magnets that as the distance increased the attraction decreased.

To test this decrease, the Curator went up, with scales, string and weights in hand, seventy-one feet, above the roof of a neighboring building, in Westminster Abbey. Hooke balanced his scales with about seventy feet of string attached to a small weight in one pan. The weight was then lowered. Hooke noted that it then took about ten grains more to balance

[6] Gunther, Vol. 6, 3/21/1666.

the scale, i.e., there was a slight increase in weight. On repeated trials, however, no change was discernible. Hooke concluded that the increase in weight was due to moisture wetting the string.

To double check his results, he moved to another position in the Abbey. In the course of repeating the experiment he found no sensible alteration in the equilibrium of the scales. This convinced him "that the first alteration proceeded from some other accident, and not from the differing gravity of the same body." [7]

Hooke, however, did not regard his negative results as decisive. Two years later, he still considered the magnetical theory a real possibility. During the course of a discussion on universal measures of time, Hooke criticized the use of a pendulum clock because, for one thing, the earth's gravity may alter. One could almost count upon this happening, claimed Hooke, because (a) all bodies are constantly changing, (b) magnetical properties alter in time and (c) if the earth is like a magnet, the poles will be more attractive than the equator.[8]

Also, in his *Micrographia,* published in 1665, under the heading of Observation LX, "Of the Moon," Hooke again mentioned the theory by way of implication. He claimed that the moon, like the earth, had a principle of gravitation. This he felt was proven by the evenness and roundness of the moon's surface. This fact, continued Hooke, opened the way for questions about the cause of the moon's gravitation. Hooke did not state definitely what he thought the cause was. However, he did rule out the possibility that it was due to the rotation of the moon since the moon did not rotate.[9]

In that same year, 1665, Hooke again attempted to solve the riddle of gravitation, but this time in a different direction. On 28 June the Royal Society adjourned *sine die* because of the plague that was sweeping London. Hooke, and others, retired to the countryside in Surrey. While there, Hooke took the opportunity to continue his work, in a well, on the magnetical theory. In a letter from Hooke to Boyle dated 3 February 1666, written after Hooke had returned to London, Hooke regretted that several of his previous letters had been lost. The Curator then went on to summarize the results of his underground experiments. Hooke found the air at the bottom of the well to be hot while the air at the top was cold. Glasses pulled up from the bottom of the well were covered with dew even though the hygroscope showed the air to be dry at the bottom. Also, candles went

[7] Gunther, Vol. 6, 12/31/1662.

[8] See Gunther, Vol. 6, 12/14/1664.

[9] See M., pp. 242-246.

out when lowered far enough. In addition, and most importantly, he could find no difference in the weight of a lowered body.

The next month Hooke reported his findings to the Society. Hooke pointed out the importance of his experiments, reminded his audience of his failure to find a decrease of weight above ground, and remarked that he realized that many notable persons maintained the magnetical theory. Hooke told the Society how he had, in the area of Banstead Downs, come upon several deep wells which afforded an ideal opportunity to test the magnetic theory.

Hooke's first set of experiments was in a well about ninety feet deep. His procedure was to weigh bodies of brass, wood, and flint both at the top and at the bottom of the well in order to note how much less the objects weighed at the bottom compared with their surface weight. His second set of attempts was in a well allowing a 330 foot descent and utilizing the same procedure as in the less deep well. The results in both sets of trials were negative. He could not help but conclude that the magnetic theory "how probable soever it might seem to Gilbert, Verulam and divers other learned men, is not at all favoured by the experiments made in these wells."[10]

Hooke goes on to emphasize the need for more accurate measuring devices. He recommended one of two instruments which he thought better than scales. One was a pendulum clock enclosed in a glass case to prevent moisture and air currents from entering, while the other was a simple device of his own design employing a small weight suspended at the end of a long arm that would easily move with the slightest change in weight.

In that same month, March of 1666, Hooke proposed an experiment, which he did not consider crucial, with a magnet and a pair of boxed scales designed to test the magnetic theory in another way. The theory behind the experiment was simple but its implementation impractical. As I interpret Hooke's plan, a body on a balanced pair of scales would be placed a certain distance from a large magnet of a certain diameter. Then, assuming the earth to be a large magnet, another body on another pair of scales would be raised from the earth's surface by a proportional amount, so as to be as far from the earth relative to the earth's size as the first body was from the magnet relative to the magnet's size. If the scales turned by the same amount, thought Hooke, one could suppose that the cause was the same, namely, magnetical attraction. The experiment, of course, could not work and apparently, since he does not seem to have tried it, Hooke realized this. To get the required distance off the earth was impossible in

[10] Gunther, Vol. 6, 3/21/1666.

Hooke's day. And, if the earth were a magnet, it would pull on the first body along with the test magnet thus throwing off any expected proportionality.[11]

Years after, however, the Curator was still toying with the notion of magnetism as an explanation of gravity. In 1674 he tried an experiment to see if an iron ring could be made to encircle a magnet at equal distances. He thought this might, if it worked, be used to explain the rings around Saturn. Again Hooke failed to get the anticipated results.[12] After this, Hooke no longer attempted to explain gravity in terms of magnetism. Since various experiments, explained above, in which certain results should have been forthcoming if the magnetic theory were true, had failed to produce the expected results, Hooke abandoned the magnetic theory.

Simultaneously with his experiments on magnetism as a possible explanation for gravity, Hooke was developing another theory which, in the future, was to be his final thought on the subject. As with the development of his thinking on the magnetic view, the evolution of Hooke's thinking on the vibratory view is seen in his reports to the Society and in one place in the *Micrographia*. The culmination of this view is seen in *The Posthumous Works* in a paper which was also originally read as a Society report. Let us, then, trace out his work chronologically.

In 1663 the Curator noted, without comment, that vibrations were produced in a glass around whose edge a wet finger is drawn. Similar vibrations could be produced in candlesticks and heated glass objects.[13] Later, in the *Micrographia*, while discussing what future discoveries his explanation of capillary action might lead to, Hooke noted that possibly the whole globe of the earth may be enclosed in a very subtle fluid, different from either earth, water, or air. This fluid, the ether, might pervade everything and somehow, Hooke does not say exactly how, be responsible for gravitation.[14]

Six years later, several experiments were performed by the Society to entertain two visiting Florentine noblemen. Among the experiments was one in which flour, and also lead shot, could be seen to ebb and flow in a vibrating glass receiver. On this occasion, Hooke expressed the view that what was observed might have something to do with gravity and that considerable other things in natural philosophy might depend upon it.[15]

[11] See Gunther, Vol. 6, 3/28/1666.
[12] See Gunther, Vol. 7, 4/30/1674.
[13] See Gunther, Vol. 6, 5/20/1663.
[14] See M., p. 22.
[15] See Gunther, Vol. 6, 3/9/1671.

Hooke was fascinated by the "flowing flour" and repeated the experiment many times. A few weeks later, he again reported upon his work. He noticed, he said, that as the vibrations of the glass increased in frequency, the flour ebbed and flowed more rapidly. Also, it appeared that the flour moved away from the point of the vibration-causing rubbings. In addition, the flour would stay in motion as long as the rubbing of the glass edge continued. Hooke felt all of this important to the understanding of natural phenomena although again he could not say exactly why.[16] This was in 1671.

About this same time we find Hooke again mentioning gravity in another report to the Society. It was well known in Hooke's day that one sure way to silence the anticopernicans was to discover an angle of parallax for a fixed star. This was a crucial experiment which, if such an angle could be observed, would prove the earth circumvented the sun. During the latter half of 1669 Hooke was endeavoring to discover an angle of parallax for the star *y* Draconis. Although he reported on his work in 1670, his report was not published until 1674, and later became the first of his *Cutlerian Lectures* (1679).

At the very end of his treatise attempting to prove the yearly motion of the earth, the Curator states that in the future he would explain a system of the world differing from any then known and he would do this without recourse to any kind of mysterious forces, psychic powers, etc. This new system, he continues, depends upon three suppositions: (1) All celestial bodies exert a "gravitating power" toward their centers. Consequently, all celestial bodies which are relatively close to one another, and this would apply to all within our solar system, attract each other to some degree. (2) The principle of inertia, stated by Hooke without proof or apology as if there was naught to dispute about it. (3) The degrees of attraction of one body for another diminish as the other body is moved farther away from the first. We will say more about this third point later.

Although Hooke does not discuss here the cause of gravity, this report is important because it shows us that as early as 1670 the Curator recognized the universal character of gravity and that any two bodies in the universe will gravitate toward one another.

This is a significant feather in Hooke's scientific cap. With the exception of Newton, who may or may not have held the same belief in that particular year, Hooke was the only thinker living at that time that we know of who so clearly and directly enunciated the principle of the universality of gravitation. That the planets were attracted by the sun

[16] See Gunther, Vol. 6, 3/30/1671.

(given the heliocentric theory) was a widely accepted view at the time. In 1666 Borelli and Hooke could regard it as elementary. But the idea that every body in the universe attracts every other body to some degree was revolutionary. Hooke himself appears to have been aware of this newness since he expressly presented his view as part of a new system. We might also point out here that Hooke was also a trail-blazer with respect to his vibratory theory of gravity. Although it is not accepted today as true, and although he did not enunciate it clearly until some years after his 1670 work, it was nevertheless an original piece of thinking on his part.

By 1682 Hooke had explicitly adopted the vibratory theory. In his "Discourse of the Nature of Comets. Read at the Meetings of the Royal Society soon after Michaelmas 1682" (i.e., October 25), Hooke had occasion to expound upon a full-blown theory of the cause of gravitation. In the course of trying to explain comets, the Curator found it necessary to discuss light and gravity. Light and gravity, he says, are the two great and universal phenomena of nature. Light is the first regular motion extending itself almost instantaneously throughout the whole universe.[17]

Referring to the days of *Genesis*, Hooke reasons that God's saying that there should be a firmament divided from the waters means the formation of the heavenly bodies in the ocean of ether. This signifies for Hooke the introduction of the second grand rule of nature, gravity. Everything, he claims, has some degree of light and gravity.[18] But what is gravity? Hooke's answer is a summary of the phenomenological data. Gravity is a name given to that power or force which makes bodies move toward one another until they unite. This is what is observed when a body falls and "unites" with the earth.[19] Now the problem is to explain what is observed.

Hooke finds all past opinions about the cause of gravitation erroneous. About the only thing that can be gleaned from them is the general agreement that bodies gravitate to the center of the earth and that the power of gravity extends some distance above the surface of the earth.[20] In addition, the power of gravity, although it probably does have a sensible limit at some great distance, acts regularly throughout the whole world and with varying degrees of force at various distances.[21]

[17] See P.W., p. 175.
[18] *Loc. cit.*
[19] P.W., p. 176.
[20] See P.W., p. 177.
[21] See P.W., p. 178.

After listing and rejecting the views of others and giving a summary of the various phenomena always associated with gravity, Hooke states his own conclusion:

> Suppose then that there is in the Ball of the Earth such a Motion, as I, for distinction sake, will call a Globular Motion, whereby all the Parts thereof have a Vibration towards and fromwards the Center, or of Expansion and Contraction; and that this vibrative Motion is very short and very quick, as it is in all very hard and very compact Bodies: That this vibrative Motion does communicate or produce a Motion in a certain Part of the Aether, which is interspersed between these solid vibrating Parts; which communicated Motion does cause this interspersed Fluid to vibrate every way in *Orbem,* from and towards the Center, in Lines radiating from the same. By which radiating Vibration of this exceeding Fluid, and yet exceeding dense Matter, not only all the Parts of the Earth are carried or forced down towards the Center; but the Motion being continued into the Aether, interspersed between the Air and other kinds of Fluids, it causeth those also to have a tendency towards the Center; and much more any sensible Body whatsoever, that is anywhere placed in the Air, or above it, though at a vast Distance.[22]

According to Hooke's way of explaining gravity, there is no need to postulate any kind of "attraction" either in the sense of a magnetic pull or as some kind of animistic appetite. Such language may have a metaphorical meaning in colloquial speech but it can have no scientific meaning. Rather, the cause – the real, concrete, physical cause – is nothing more or less than a series of rapid vibrations starting in dense bodies and from thence being communicated to the ether. Moreover, because the universe contains numerous dense bodies (stars, planets, satellites) each similarly vibrating and because the ether extends throughout the whole universe, the phenomenon of gravitation is also universal. It would seem, in addition, that *any* dense body, such as a small rock, also exerts a gravitational "pull" as does some huge rock (e.g., a planet). However, given the minuscule effect such a small body would have when compared with that of a huge body in whose vicinity such a small body would be, Hooke could afford to ignore this consequence of his theory.

If someone should think it strange that an outward vibration could make something come down, Hooke suggests that they observe a tradesman driving a hammer-head onto the helve by hitting the upright end of the helve thus causing the head to ascend. Hooke also uses the analogies of vibrating bells and water or flour in a glass to make his theory comprehensible. We do not see the metal in a struck bell vibrating but we "hear" the vibrations as they are conveyed to us through the air.

[22] P.W., pp. 184-185.

Likewise, just as vibrations in the earth set up vibrations in the ether, so vibrations in a glass container will set up vibrations in the glass' contents.

In order to have his view conform to the law of falling bodies, Hooke supposes a long series of compounded vibrations. Hooke claims that for each stroke of the vibrating globe or medium, one degree of velocity of descent is given to a heavy body. The power of gravity, therefore, will vary with the frequency of the vibrations. Now assume, suggests Hooke, 1000 pulses a second. If this remains constant, then a body would receive equal degrees of acceleration in equal times so that the total distance fallen (provided that you had a body that was very dense relative to the medium) would be proportional to the time squared, just as Galileo had demonstrated. And, if the second of time were subdivided into a thousand parts, the body would still receive one degree of acceleration in each moment. The result is a compounded acceleration actually noted in falling bodies.[23]

Some time between 1682 and 1684 (the exact date is unknown) Hooke wrote a brief summary of his views on magnetism which Waller inserted after his treatise on comets and gravity. We find there an interpretation of magnetism in terms of his vibratory view. The magnetic power of the earth is due to the vibration of its parts from north to south and *vice versa*. The medium for the vibrations is the ether. Loadstones, claims Hooke, are highly dense substances capable of picking up these vibrations.[24]

### *5.3 The Mathematics of Gravitation.*

We have seen how Hooke explained gravity and elliptical paths taken by the planets around the sun.

It now remains for us to consider the inverse square law of universal gravitation. The degree of attraction of the sun for the planets appears to have been a problem for Hooke from about 1670 to 1679, since we find the problem mentioned for the first time at the end of his 1670 treatise on the yearly motion of the earth. Between 1670 and 1679 we find no record of Hooke's thoughts on the subject. In 1679, however, the Curator engaged in a series of letters with Newton. It is in these letters, and in a letter from Aubrey to Wood, that we find Hooke's answer to the problem. One finds in these letters a revelation of Hooke's thought drawing an answer from Newton which in turn elicited further remarks from Hooke, and so on. We must therefore follow in chronological order what Hooke had to say

[23] See P.W., p. 185.
[24] See P.W., pp. 192, 481-483.

on the problem at hand from 1670 to its solution, as far as he was concerned, in 1679.

We have already mentioned Hooke's 1670 (published in 1674) treatise attempting to prove the earth's yearly motion in which he claimed to have discovered an angle of parallax for a fixed star. At the end of his treatise, Hooke promised to give to the world an explanation of celestial motions resting upon three suppositions. The third supposition was that the power of gravity between two bodies would diminish by various degrees as one was removed farther and farther from the other. With respect to this third supposition, Hooke admits in 1674 that "what these several degrees are I have not yet experimentally verified." Nevertheless, he promises to work on the problem in the future and urges others to do so also. The Curator concludes his treatise with the statement that "the true understanding thereof will be the true perfection of Astronomy." It is clear that as late as 1674 Hooke did not have the problem of the relationship between distance and force of gravity solved. The hope of solving the problem in the near future may have been the reason for his four year delay in publishing his earlier report.

The period between 1674 and 1679 appears to have been a crucial one with respect to Hooke's views on the inverse square law. We know that he did not have the law in 1674 but that he did have it by the end of 1679. We can perhaps narrow down the crucial time period to under a year in length. In 1678 Hooke published a treatise entitled *Cometa* in which he presented his carefully made observations upon the bright comet of 1677. One would expect that if Hooke had knowledge of the inverse square law he would have mentioned it at that time. Comets and celestial mechanics were topics which always seemed to go together for Hooke. When mentioning one subject, he would usually bring in the other. From what he said in 1674, it appears that he planned to spend a good deal of time thinking about the matter and his *Cometa* would have been an ideal occasion to mention the fruits of his labors. But, he did not.

It would seem, then, that some time between 1678 and 1679 the Curator came upon the inverse square law, at least as a hypothesis. That Hooke finally did grasp the inverse square law is evidenced by his correspondence with Newton. Between 1679 and 1680 seven letters passed between Hooke and Newton. Between 1686 and 1687 a series of letters passed between Newton and Halley. These are devoted largely to a discussion of publication details and what role others actually had in the composition of Newton's *Principia.* On June 2, 1686, the Royal Society authorized the publication of the *Principia* and put Halley in charge of the necessary arrangements. Halley also paid for its publication. After its publication, Hooke publicly attacked Newton for treating him unfairly. Hooke based much of his case on the 1679-1680 correspondence with Newton. From Halley's point of view, the 1686-1687 Halley-Newton correspondence was mainly a case of attempting to soothe Newton's hurt feelings. Since Hooke and Newton were no longer on speaking terms, Halley found himself playing the role of a mediator.

The trouble began with a 24 November 1679 letter from Hooke to Newton.[25] Hooke, in his capacity as secretary to the Royal Society, wrote to Newton reminding him that his correspondence with the Society was lagging and asking him to forgive and forget any enmity there may have been between them. As a sign of his desire to be on friendly terms with Newton, Hooke asked the Cambridge mathematician to comment upon any of his hypotheses or opinions, and especially upon his view that celestial motions are compounded of a "direct motion tangent and an attractive motion towards the central body." Newton replied on the 28th of November, expressing his thanks for Hooke's kindness and saying that he would like to keep up his philosophical correspondence but that at the moment his interests lay elsewhere, mainly in "country affairs." Nevertheless, he went on to say "I shall communicate to you a fancy of my own" and proceeded to present his idea on the path of a falling body on a moving earth. After his presentation, Newton again expressed his opinion, or rather, lack of opinion, on Hooke's hypothesis.

Concerning the trajectory of a falling body, Newton suggested that the path would be a spiral line deviating *to the east* as the object approached the center of the earth. This, he thought, would prove the diurnal motion of the earth. An ancient objection against a rotating earth was the "fact" that a body thrown straight up comes down over the same spot. This was interpreted to mean that the earth did not move out from underneath the body while it was unattached to the earth's surface. Assuming the diurnal motion, as long as a body is anchored on the earth's surface, it will rotate at the same speed (about .3 miles per second) as that surface. But the same body off the ground, it was commonly thought, would quickly fall behind its original point of connection with the earth. Where would such a body land? To the west, of course. But this does not happen. Therefore the earth does not move. If it could be shown that a body does not come down over the same spot, this ancient objection could be overcome.

Hooke wrote his answer to Newton on 9 December 1679 and read the

[25] More has reprinted with commentary this whole series of letters in his life of Newton.

same before a meeting of the Society on December 11th. Rather than a spiral to the east, claimed Hooke, a body would fall to the S.S.E. Furthermore, the path the body would follow, supposing the earth permeable, would be a flattened ellipse forever moving about the earth's center if there were no medium and an eccentric elliptical spiral gradually closing in on the earth's center if there were a medium. It is significant that he realized that a falling object might possibly never fall to center or, if it did, it would circle the center of the earth many times before hitting its mark.

Newton answered almost immediately. In his letter of 13 December 1679, Newton attempted to rectify his earlier error, but without success. Hooke, apparently unaware of the mental anguish he was causing Newton, publicly read Newton's answer on the 18th of December. At that time he also reported to the Society that he had tried the suggested experiment three times with positive results in each case. At the beginning of the following year, Hooke reported that he had dropped a weight twenty-seven feet into a pan of clay, marked with lines to show the position of the weight when hung on the perpendicular, and noted that it had stuck in the S.E. quadrant. The Curator was told to repeat the experiment with witnesses.[26]

[26] See Gunther, Vol. 7, 1/22/1680. See also Hooke's *Diary* for 16 January 1680 and his letter to Newton on 17 January 1680 for statements of his positive results indoors. For a study of the problem before Hooke see A. Koyré, "A Documentary History of Fall from Kepler to Newton," *Transactions of the American Philosophical Society,* Vol. 45 (1955), part 4, pp. 329-395. For a study of the topic after Hooke see A. Armitage, "The Deviation of Falling Bodies," *Annals of Science,* Vol. 5 (1947), pp. 342-351. The Curator's predecessor with respect to the body's path to the earth's center was Borelli, while it was Newton who urged Hooke to provide experimental verification of the earth's diurnal motion. According to Borelli, a body would fall in a curved path, moving to the east of the point from which dropped, on its way to the earth's center. To understand what he had in mind one must imagine a long hollow tube extending from the earth's center to its surface on the equator. This tube will of course rotate with the earth. Now imagine a uniformly accelerated stone descending in the tube. The stone has two motions: one down and one east. If exaggerated and graphed, the total path would look like the cross section of a snail's shell. However, the actual deviation from perpendicular would be very small. Under ideal conditions in the twenty minutes it takes a stone to fall the 4,000 miles to the center, the earth would have moved a mere 330 miles (about five degrees of arc), thus inscribing a path so close to the perpendicular, especially when near the earth's surface, as to be indistinguishable from it. On 15 June 1668, James Gregory, who had studied in Italy, reported upon the work of Borelli and others to the Royal Society. Hooke was familiar with most of what Borelli had to say. At a meeting of the Royal Society in 1680, Hooke mentioned that he had followed Borelli's work with interest and was sorry to hear that he had died. See Gunther, Vol. 7, 8/9/1680. We might also mention that,

DIAGRAMS ILLUSTRATING HOOKE'S
VIEWS ON THE TRAJECTORY OF FALLING BODIES

without
a medium

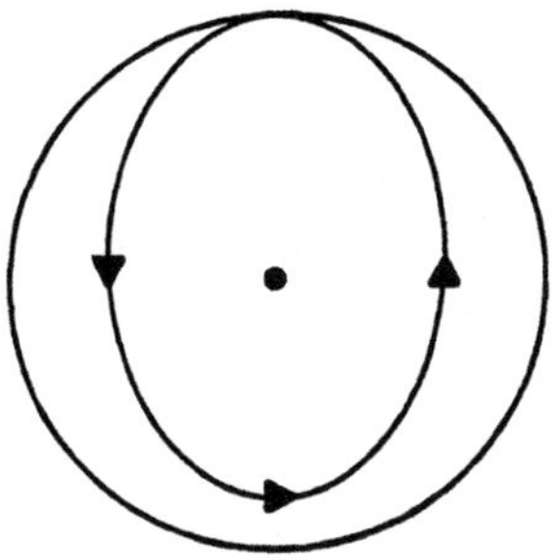

with a
medium

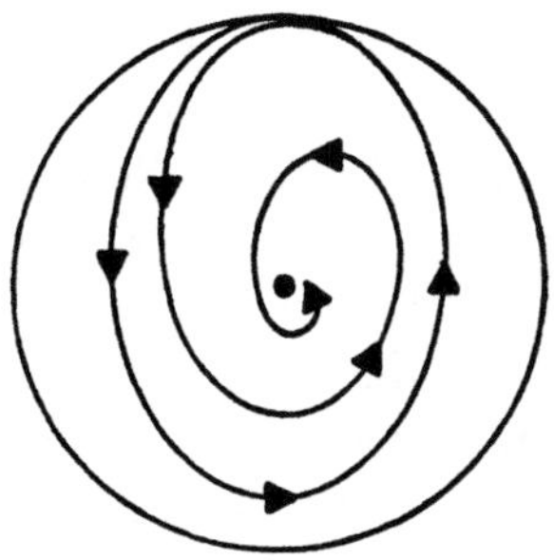

In his reply to Newton, Hooke seems oblivious to the fact that a deep personal rift had opened between Newton and himself. Hooke wrote to Newton on 6 January 1680 stating his criticism of Newton's view on the trajectory of a falling body and also reporting to Newton his experimental work on the problem.

In the same letter Hooke makes the statement which was later to be such a bone of contention between the two men. In reference to his own view concerning the relationship between a falling body and gravitational

theoretically, Hooke's results were correct. However, practically speaking, he could never have gotten the results claimed. Given the latitude of London and the low heights from which he worked, the deviations from the perpendicular would be imperceptible. This indicates that he had not read Borelli very well. In fact, it is highly doubtful that Hooke was even acquainted with Borelli's *Risposta di Gio* (Messina, 1668) in which Borelli showed that the deviation of a freely falling body from the perpendicular would be so small as to be insensible. Undoubtedly, Hooke, and those witnessing the experiment, did see the weight fall into the S. S. E. section of the pan of clay. But how can this be reconciled with the fact that we know from modern calculations and fine, precision instruments that such a thing could not have been observed? We can only call upon the crudity of their apparatus and the anxiousness of their mental state to see the results seen.

pull he states that "my supposition is that the Attraction always is in a duplicate proportion to the Distance from the center Reciprocall." Also, when dealing with the vast expanse of space involved in celestial mechanics, thought Hooke, one could measure distances from the centers of the bodies affected.

Newton refused to answer Hooke's letter. Hooke, however, apparently still unaware that he had offended Newton, wrote again on 17 January 1680. He again told Newton about his work on the paths of falling bodies and, in addition, requested Newton to employ the calculus in order to help prove the inverse square law of gravitation.

Newton did not answer until December of 1680. Even then his answer was buried in a context which had nothing to do with the topics of their previous letters. An Italian medical doctor had requested Newton to intervene for him with the Society in order to gain permission for him to dedicate his book to the Society. Newton forwarded his request to Hooke. In the course of doing so, Newton adds to his 3 December 1680 letter: "For the trials you made of an experiment suggested by me about falling bodies, I am indebted to you thanks which I thought to have returned by word of mouth, but not having yet the opportunity must be content to do it by letter." Thus ended the correspondence between Hooke and Newton, concerned with gravity.

Upon termination of their correspondence, the two men went their own ways, seemingly unconcerned about what the other was doing. Hooke published his views on the causes of light and gravity. Newton, meanwhile, was proceeding with his calculations on "Hooke's Hypothesis." In May of 1684 Halley consulted Newton at Cambridge on his progress. Newton responded by sending Halley a copy of his *De Motu*, the nucleus of what was to be the first book of his *Principia*. On 10 December of that same year, Halley reported to the Royal Society upon the contents of Newton's treatise. In February of 1685, Newton sent Aston, a secretary to the Society, a completed copy of his *De Motu*.

It would seem, however, that as late as September 1685 Newton still had his doubts about the validity of the inverse square law. We know from a letter written by Newton to Flamsteed, the Royal Astronomer, dated 19 September 1685, that Newton was grateful for some recent figures on observations Flamsteed had sent him. These figures, claimed Newton, removed the doubts he had about the law because, as he had previously thought before receiving Flamsteed's calculations, the influence of Jupiter and Saturn upon the other planets did not seem as great as the inverse square law supposed they would be.

Finally, on 21 April 1686, Halley announced to the Society that Newton's book was ready and, on the 28th, Dr. Vincent presented the manuscript of book I of the *Principia* to the group. It was at this point that Hooke created a stir, condemning Newton for having stolen his ideas and for having failed to so much as mention him in any part of the work. Halley took upon himself the unpleasant task of reporting Hooke's remarks to Newton, which he did in a letter to Newton dated 22 May 1686. After a month's deliberation, Newton sent Halley a long letter on 20 June 1686. Newton condemned Hooke as a liar and attempted to present his side of the story. Halley replied on the 29th with a very sympathetic letter in which he more or less agreed with Newton. His main point, however, was to emphasize that Hooke had not published before Newton and, therefore, must waive any claim to priority. The debate thereafter subsided and Newton, under gentle pressure from Halley, finally agreed to at least mention Hooke, along with Wren and Halley, in his work.

While on the subject of correspondence, one more important letter should be mentioned. In support of Hooke's claim to priority, Hooke's friend Aubrey wrote Anthony Wood (1632-1695) attempting to give him the background to the debate.[27] Wood had previously published a work entitled *The History and Antiquities of the University of Oxford* (1674) in which he had neglected to mention either Hooke or Aubrey.[28] In order to make sure that such a thing did not happen again, Aubrey, after the publication of Newton's *Principia*, took it upon himself to compose a letter to be read by Wood. Before he sent the letter to Wood, however, he gave it to Hooke so that he might add or delete anything he cared to. The finished Aubrey-Hooke letter, dated 15 September 1689, begins by quoting *in extenso* the end of Hooke's 1674 treatise on the proof for the motion of the earth.[29] In the middle of Hooke's words concerning his third supposition Aubrey interjects in parentheses a sentence of his own following Hooke's line "I have not yet experimentally verified," in which he recalls how in 1678, Hooke related to Newton "these degrees and proportions of the power of attraction in the coelestiall bodys and motions" by means of a series of letters. Other than the erroneous date (it was in 1679 not 1678 that Hooke wrote to Newton), what Aubrey says is a matter of record. At the conclusion of his quote from Hooke, Aubrey goes on to

[27] Concerning the close friendship between Hooke and Aubrey see M. 'Espinasse, *Robert Hooke* (London, 1956), ch. 6.

[28] For the background to this letter see More, p. 353. Herivel does not mention it.

[29] Hooke, in his *Diary* for 15 September, records laconically "letter for Mr. Wood about Newton."

summarize the pertinent contents of these letters, pointing out that Hooke revealed to Newton the "whole" hypothesis, namely, the "gravitation was reciprocall to the square of the distance, . . ." At this point Hooke adds a line of his own which reads as follows:

> which would move the motion in an ellipsis, in one of whose foci, the sun being placed, the aphelion and parhelion of the planet would be opposite to each other in the same line, which is the whole coelestial theory, concerning which Mr. Newton hath a demonstration, . . .

Aubrey's narrative then continues: "not at all owning, he receiv'd the first intimation of it from Mr. Hooke." The remainder of the letter is a disparagement of Newton relative to Hooke, and ends with an exhortation to Wood to read Hooke and "doe him right." As far as Aubrey could see, the case was closed and Hooke had won. Wood, however, was not so impressed. The first edition of his *Athenae Oxonienses* (London, 1692) failed to include a biography of Hooke, while the second edition, written about 1694 but not published until 1721, included only about a page on Hooke and in no way exalted the Curator over the Cambridge mathematician.

After discussing Hooke's work in astronomy, one cannot avoid probing deeper into Hooke's relationship to Newton. To engage in an *exhaustive* discussion of this topic would take us much too far afield to be contemplated here. We can, nonetheless, mention several points on Hooke's side and several on Newton's side which any fully developed presentation must take into consideration.

The solution to the Hooke-Newton debate over who should be put first in the history books will hinge upon the answers to three questions: (1) Did Hooke work out the mathematical proof for the theory of universal gravitation independently of Newton, thus entitling him to an equal place in the history of Western thought in this respect? (2) Did Newton have at least a non-mathematical knowledge of the theory previous to his correspondence with Hooke, thus freeing him from any indebtedness to Hooke in essentials? (3) Is the statement of the key principles needed in the resolution of this problem, without a rigorous mathematical proof, sufficient to justify a claim to priority? With respect to the first question, on Hooke's side, we must note Patterson's speculations.

Louise D. Patterson has expressed the opinion that Hooke has been poorly treated by historians. As the result of misinformation passed on by Newton and his friends, our present-day appraisal of the Curator is not what it should be. As Patterson puts it:

Although the gravitation theory is generally considered the most important product of seventeenth-century science, the evolution of the theory prior to the publication of Newton's *Principia* in 1687 has been curiously neglected by the historians. By reference to one volume after another dealing with that period it may be seen that a stereotyped account of the history of the theory based chiefly upon the reminiscenses of Newton and his editor, Edmund Halley, and seldom referred to other seventeenth-century sources, has been transmitted from author to author with little variation: What Whewell and Wheatley have called "the Baconian period" of the Royal Society has been largely overlooked, except as it provides material for brief comment about the precursors whose failure to perfect the gravitation theory adds lustre to Newton's achievement.[30]

If one were to look into Waller's biography of Hooke prefixed to *The Posthumous Works,* one would not find any reference to the Hooke-Newton argument over who was to be given credit for the theory of universal gravitation. Perhaps Waller's omission merely reflected the fact that in 1704 there was little interest in a debate which had taken place fifteen years before. Then again, thinks Patterson, the situation might not be quite so simple.

As mentioned in Chapter I, toward the end of his life, beginning around 1683, Hooke became lax in preparing his papers for publication. Waller presented this as an unfortunate state of affairs and said that, although he would like to give a more complete account of Hooke's work during these later times, he decided instead not even to mention the titles of the various papers since this would but create an uneasy curiosity in the reader without any satisfaction. These could not have been, reasons Patterson, the papers published later by Derham since this would not explain Waller's remark about uneasiness and, besides, if they were, they could have been handled by Waller himself. In Birch, one finds only those papers made public before the Royal Society during that period of time. Patterson postulates the following as possibly having been the true chain of events: Waller, first of all, received from Hooke's niece, along with the works that were later published, certain private papers of Hooke's which showed Hooke to definitely have been the true author of the theory of universal gravitation. Waller, however, then secretary to the Royal Society, did not make these papers public due mainly to the fact that Newton was then president of the Royal Society and held in very high regard. Instead, he held the papers and, shortly before he died, gave them over to some unnamed party (or parties) who was to both preserve the papers and, after Newton's death, to see to their publication. Unknowingly,

[30] L. D. Patterson, "Hooke's Gravitation Theory and Its Influence on Newton," *Isis,* Vol. 40 (1949), pp. 327-328.

thinks Patterson, the papers were given to someone who was a better friend to Newton than he was to truth. The papers, of course, have since totally disappeared.[31]

As for Newton's side to this first issue, one need only point out that Patterson's case is circumstantial, at best. Even granting the possible defects in Newton's character and personality as depicted by her,[32] the fact remains that he was a mathematical genius. There is nothing extant in Hooke comparable to the works of Ptolemy, Copernicus, Kepler, or Newton. If Hooke did work out a mathematical proof for the theory of universal gravitation, and if it was deliberately kept from public view, it may some day come to light. However, until such evidence is forthcoming one must consider it wiser to judge Patterson's claims against the background of what is definitely known about Hooke's work and ability. Doing this leads us to suppose that he *probably* did not accomplish the mathematical feat in question.

Turning now to the second issue, it is clear from Newton's own words that he himself did not feel any indebtedness to Hooke since he himself had thought of the essential notions communicated to him by Hooke in 1679 well before that date. As the result of an interview with Newton on the subject in 1694, William Whiston (1667-1752), who was appointed by Newton to fill his Lucasian Professorship in mathematics and astronomy at Cambridge upon Newton's resignation in 1701, reported that, according to Newton, Newton had hit upon the theory many years before while still a young man.[33] Henry Pemberton (1694-1771), the editor of the third edition of the *Principia,* repeats the story. Even later, about 1714, Newton himself wrote a brief, unpublished memorandum outlining his intellectual history with respect to his discovery and proof of the theory of universal gravitation. Newton did not mention Hooke as he recalled how, during the plague years of 1665 and 1666, he was aware of the inverse ratio proportion. In fact, he had even "compared the force requisite to keep the moon in her orb with the force of gravity at the surface of the earth, and found them answer pretty nearly." [34]

Patterson put little stock in these late reminiscenses of Newton's. Even

[31] See *ibid.,* pp. 328-341 for details. Hooke did say during the critical years 1684-1686 that he wanted to present the Society with a paper on celestial bodies but decided not to since Newton would shortly do so. Newton's treatise was referred to as being "now in the press." See P.W., pp. 173, 330.

[32] See *ibid.,* Vol. 41 (1950), pp. 32-45.

[33] See W. Whiston, *Memoirs of the Life of Mr. William Whiston by Himself* (London, 1749), Vol. I, pp. 35-38.

[34] As quoted by More, p. 290. See also Herivel, pp. 66-67.

L. T. More, who takes the role of Newton's champion in his biography of Newton which, since 1934, has become a standard biography of Newton, must admit that there is something a little strange about these recollections on Newton's part. With respect to the early work of Newton during 1665 and 1666, More remarks that the "straightforwardness" of the tradition about Newton is only "apparent." More cannot understand why Newton was not satisfied with a result that was very close to the ideal result. After all, More points out, according to his own words Newton himself at that time expected his calculation to be an approximation. In fact, he had deliberately introduced various simplifications of the data in order to facilitate his work. More, nevertheless, accepts "the tradition" as true, accounting for the apparent discrepancies by assuming that Newton was distracted from his work on gravitation by other interests, primarily optics.[35] More might also have mentioned that it is very strange that Newton did not mention explicitly and exactly this early work in his 20 June 1686 letter to Halley answering Hooke's charges since that would have been the most likely place for Newton to have done so.

More recently, new light has been thrown upon this issue by John Herivel.[36] This commentator points out that in Newton's correspondence with Halley following Hooke's attack, Newton would most certainly have been looking for *dated* evidence that he had knowledge of the theory prior to 1679. Such, however, did not exist (and still does not) thus causing Newton to neglect mentioning any such work he may have done during the plague years in his 20 June 1686 letter to Halley.

Newton did, however, mention to Halley in his 20 June 1686 letter that he had written to Huygens via Oldenburg on 23 June 1673 stating some of Hooke's key notions and that he had also composed a short treatise on circular motion "some time before I had any correspondence with Mr. Oldenburg, and that's above fifteen years ago." In his 14 July 1686 letter to Halley he claimed, more precisely, that the short treatise was written eighteen or nineteen years before.

What must be seriously considered by anyone undertaking a thorough study of the issue now under discussion is that, although the part of Newton's letter to Huygens via Oldenburg supporting his claims against Hooke was and is missing, Herivel (following A. R. Hall's original discovery in 1957 of the short treatise on circular motion) thinks this work to be the one referred to by Newton in his 1686 correspondence. Hall and Herivel admit that they cannot be absolutely certain of it but,

[35] See More, p. 291.
[36] See Herivel, pp. 72-76.

if such is the case, it would definitely place Newton in possession of Hooke's key notions on celestial mechanics before the year 1670.[37]

Passing on now to our third question, and assuming, for the sake of argument, that Newton did learn the basic ingredients of universal gravitation from Hooke, we may distinguish two sides or aspects with respect to the nature of the debate in question. One aspect of the case is that of the mere statement of the principles involved in the resolution of this problem in celestial mechanics. The other side of the case is the production of the actual mathematical proof needed to substantiate the claims put forward in the statement of the essential principles. If one regards the statement of the key principles needed in the resolution of some problem of this type as sufficient to justify a claim to priority, then the palm must go to Hooke insofar as he gave to Newton material necessary to Newton's work. This was Hooke's way of looking at the issue as can be clearly seen from the words he interpolated into Aubrey's letter to Wood quoted above. If, however, one regards the former alternative as insufficient without the mathematical verification, then Newton carried the day. This was Newton's way of looking at the debate as can be clearly seen from his 20 June 1686 letter to Edmund Halley. Even if he could not present Halley with an airtight case against Hooke based upon dated documents, he still did much more than Hooke ever did. Halley had written Newton asking him to include in his work some sort of printed recognition of Hooke's contribution, perhaps a few lines in the Preface. Newton answered in unequivocal terms that he had no intention of doing so and resented even being asked by his friend Halley to do so. Newton argued that just as Kepler had only guessed that the planetary paths were elliptical but did not prove it mathematically so Hooke had but a vague notion of universal gravitation based upon some sort of guess or hunch. Inventing hypotheses is one thing; proving them is quite another. In fact, continues Newton, "There is so strong an objection against the accurateness of this proportion, that without my demonstrations, to which Mr. Hooke is yet a stranger, it cannot be believed by a judicious phi-

[37] See Herivel, pp. 192-198. Although it does not directly discuss the question of priority, an article by Florian Cajori should be mentioned here. Cajori maintains that the twenty year delay in publishing the inverse square law was not due to inaccurate calculations for the distance of one degree of latitude as is most commonly claimed, but was rather due to fears Newton had concerning the *universality* of the law, i.e., whether it could be applied to all bodies everywhere, even those very close to the earth. Implicitly, Cajori sides with those who accept Newton's claims to have been aware of the law before Hooke. See his "Newton's Twenty Years' Delay in Announcing the Law of Gravitation," *Sir Isaac Newton 1727-1927* (ed. by the History of Science Society, Baltimore, 1928), pp. 127-188.

losopher to be any where accurate." He concludes by affirming once more that Hooke can no more claim credit for the inverse square proportion than could Kepler for the ellipses; guesses do not count.

These are the key issues which must be considered in the Hooke-Newton debate over universal gravitation. A definitive resolution will come, if at all, only after many more years of research. We can, however, at least give a tentative endorsement to Butterfield's remark concerning Hooke's relationship to Newton; namely, "His reputation has risen, with the development of historical research, therefore; though the glory of Newton has not been eclipsed." [38] Rather than respecting Newton less, one must respect Hooke more. This same sentiment is echoed by Armitage and Koyré as can be seen from their pertinent works already cited. On the basis of his own research, this author can agree with this basic position taken by these noted historians of science with respect to Hooke's place in history. Although admittedly not on a par with Newton in mathematical matters, Hooke certainly far exceeded Newton in experimental matters. And, while in the process of delving into nature's secrets, he produced several brilliant insights, some of which were closely connected with the great Newtonian synthesis.

[38] H. Butterfield, *The Origins of Modern Science* (New York, 1962), p. 167. The reader may be interested in knowing what it was that Hooke was apparently incapable of doing. By combining Huygens' law for centripetal (or centrifugal) force with Hooke's hunch on universal attraction Newton was able to come upon a formula which agreed with Kepler's third law for planetary motion. This agreement between mathematical reasoning and observation was the proof to which Hooke was yet the stranger.

$$F = \frac{mv^2}{D} \quad \text{(Huygens' Law)}$$

$$V = \frac{2\pi D}{T} \quad \text{(for angular velocity)}$$

$$F = \frac{m_{(planet)}V^2}{D} = \frac{m_p}{D} \cdot \left(\frac{2\pi D}{T}\right)^2 = \frac{m_p 4\pi^2 D}{T^2}$$

$$\frac{m_p 4\pi^2 D}{T^2} = G\,\frac{m_{(sun)} m_{(planet)}}{D^x}$$

$$\frac{Gm_s m_p}{D^x} = \frac{m_p 4\pi^2}{1} \cdot \frac{D}{T^2} = \frac{Gm_s m_p}{m_p 4\pi^2} = \frac{D^x D}{T^2}$$

$$\frac{Gm_s}{4\pi^2} = \frac{D^{x+1}}{T^2}$$

$$k = \frac{D^3}{T^2} \quad \text{(the time needed for a planet to complete its orbit as related to its distance from the sun)}$$

Therefore, the distance must be squared in the formula for universal gravitation:

$$F = G\,\frac{m_s m_p}{D^2}$$

CHAPTER VI

# HOOKE'S PLACE IN THE HISTORY OF MECHANICS

Looking back into the seventeenth century, we saw the importance of Francis Bacon with respect to the attitude toward natural philosophy held by Hooke and many of his contemporaries. In his own time, Hooke, largely due to his pivotal position as Curator of experiments to the Royal Society of London, held an important place in seventeenth century English intellectual circles. He was renowned as an indefatigable worker and experimenter. Although we did not attempt to set out all of Hooke's accomplishments and attempted accomplishments in the areas other than his mechanics, much space could be devoted to these other areas. The Curator was incurably curious about everything and anything. His mechanical contrivances and work with the then fledgling microscope made him known on both sides of the Channel.

This same disposition in his character, however, also served to work against him as well as for him. There was a tendency for the Curator to be too much interested in all subjects. Consequently, he could not devote all his energies to any one restricted area of interest, such as mechanics.

Of course, there is no way of telling what would have occurred if he had devoted all of his time and talent to one topic or area of interest. This does not mean that Hooke was totally unorganized or erratic in his investigations of phenomena in nature. One can discern by carefully sifting through Hooke's treatises certain main areas of interest which Hooke followed out by a series of experiments and observations. By and large, however, a thoroughly consistent series of experiments, carried out over a long period of time and focused exclusively on one narrowly defined area is not to be found in Hooke's extant work. Hooke was a man too easily distracted by the vast variety of interesting phenomena in the world. Over and over again one bears witness to a man who is all too ready to put down what he is doing in order to take up a more interesting topic or investigation. The reason for this is implicit throughout the whole of Hooke's works; namely, there was so much to do and so little time to do it in.

With respect to his work on mechanics, we saw that Hooke's early interests lay in the nature and *modus operandi* of fluids, especially the air. At the time Hooke lived, the air was still virgin territory for scientific investigation. The Curator was interested in the air as it related to respiration and combustion. Hooke was also interested in knowing how far into the ether the air extended. Hooke thought that if he could penetrate the secrets of air he could understand other phenomena as well. Such phenomena were, for instance, the shape of oval or rounded objects, such as stones, drops of liquid, or even the heavenly bodies; the odd or distorted appearances of various terrestrial and celestial objects when seen at a distance; the rising of liquids in fine tubes to various heights depending upon the kinds of liquids and tubes used; clouds; and various types of weather. He believed this because he knew from experience that man lives in a sea of air.

As far as the nature of the air is concerned, Hooke decided that it was a collection of vapors and other light bodies suspended in the ether. The ether itself was thought to fill the whole universe, pervade all bodies, remain stagnant unless forcibly moved, and had a great deal to do with events, such as meteors, observed in the heavens.

It was while investigating the mechanics and properties of the air that Hooke came upon "Boyle's Law" through the back door, so to speak. It is unfortunate that Boyle's name became attached to the law rather than Hooke's name. About 1660 Hooke was working as Boyle's chief assistant. As we have seen, it was during this period in his career that he carried out various carefully designed experiments in order to discover the cause of capillary action and the relationship between the pressure and volume of a certain portion of air. As Boyle himself states in the work in which "Boyle's Law" appears, the proposition that gas pressure times volume is equal to a constant was stated and verified by his assistant Hooke. It is true that at least two other persons were also named by Boyle as working in the same direction as was his assistant Hooke. However, it is clear from the context of Boyle's report that it was Hooke who actually accomplished the experiments and consequent proof in question. It would not be going beyond the evidence, then, to maintain that it is really Hooke who should be given credit for having discovered "Boyle's Law."

History has proven the pressure times volume formula to have been a real contribution to the physics of gases. This does not mean that Hooke's formulation was perfect. Modifications have been made in the relationship so as to take into account the relationship of the temperature to the other two factors as well as modifications needed to account for variations from the "ideal" states of different types of gases. In a similar way, Hooke misjudged the comprehensiveness of "Hooke's Law." His Baconian trait of seeking direct, simple solutions to problems that interested him has worked against him here.

He also assumed the mechanistic world view of Descartes and proceeded to explain everything accordingly. In a world of picturable matter in motion there is no room for intrinsic powers or spirits in things. Therefore, any explanation of nature must proceed without them. The only cause or force needed to explain the existence and conservation of matter and motion is God.

Furthermore, all bodies and parts of bodies are in constant motion, i.e., local motion. The type of local motion postulated by Hooke to exist in all bodies was a vibratory motion. Descartes had maintained, in his celestial mechanics, that the most subtle material substance, the ether, was in constant motion. This motion was a swirling type and was used by Descartes to explain the movement of heavy bodies toward a center. Hooke, however, attempted to unify all of the major universal phenomena, i.e., heat, light and gravitation, by maintaining the existence of a universe in which there were, instead of vortices, countless vibrations within all material entities. Heat, he claimed, was nothing more than the rapid and violent agitation of the small parts of bodies. The degree of heat depended upon the rapidity of the vibrations. Light, also, depended upon the existence of rapid vibrations within the ether. Gravity, too, was the result of millions of vibrations, within both the ether and grosser bodies, which tended to move larger, grosser bodies down toward the center of the streams of vibrating particles. Such was Hooke's kinetic theory of matter. This postulate of Hooke's is accepted today as basically true and marks, therefore, a real contribution to mechanics on the part of Hooke. One may even go further and perhaps call Hooke the father of wave mechanics. His vibrations were regular, timed, and patterned. And, although he never gave us a picture of them, they could not have been too unlike the models used by modern physicists.

On the topic of falling bodies near the surface of the earth, we saw how Hooke and Galileo both agreed and disagreed. Hooke viewed Galileo's rule concerning the rate of descent of a falling body as an ideal case which would never be found in actuality. This was a fact ascertained by Hooke himself in a series of experiments designed to verify Galileo's contention that there would be equal increments in velocity in equal times. If an examination of another's theory or ideal experiment can be considered a contribution to some field of study, then Hooke's name must henceforth appear with that of Galileo's.

The path that a falling body would take near the surface of a moving earth posed a more difficult problem for seventeenth century thinkers to solve than that of the rate of descent. Hooke postulated that a body, if it could fall freely through the earth, would circumvent the center indefinitely (if there is no medium) or would take a long curved path, spiraling in upon itself, before it finally came to rest at the earth's center (if there is a medium). We cannot tell from Hooke's extant works *why* he believed this but it is certain that he did. This belief marks a partial contribution to mechanics. Hooke, in postulating an indefinite circling or a delayed arrival at the center, was far more advanced in this respect than any other thinker of his time before Newton.

As far as the interrelationship between moving bodies coming into contact with each other is concerned, Hooke performed several experiments in an effort to establish the existence of some constantly recurring pattern. It was his belief that any body could move any other body regardless of how large the second body was in comparison to the first body, if the first body was traveling at a high enough speed. He concluded that if the speed of a body with a constant size and density was doubled, then it could move twice as much as it could when traveling at its original velocity. Hooke also determined that in order to double the body's velocity (momentum) it would be necessary to quadruple the distance fallen (if the weight is constant) or quadruple its weight (if the distance is constant). This relationship was generalized by Hooke and can be summarized by the expression $NV_0 \longleftrightarrow N^2(D_0 \text{ or } W_0)$. This can be regarded as a development of Galileo's work on falling bodies. Also, Hooke's work may form an intermediate stage between Galileo's and Descartes' momentum formula ($F = wv$) and Newton's second law of motion ($F = ma$).

Let us not overlook Hooke's Law. Hooke enunciated his law in 1676. It was a generalization based upon his previous experiences with matter in motion and upon several simple experiments with wire and springs carried out about that time. As noted, Hooke did not bring his discovery to perfection by noticing its limitations. Nevertheless, what he discovered with respect to the relationship between stress and strain was surely a step forward in the study of nature as is indicated by the continued use of Hooke's Law even today.

With respect to Hooke's work touching the movements of the earth, several contributions and partial contributions can be listed. In Hooke's time, the heliocentric theory of planetary motion was not widely accepted. The Copernican hypothesis was widely known, but it had not been proven to be true in a manner which would exclude the possibility that some other system (e.g., the geocentric view or Tycho Brahe's) was true. Hooke, however, was attracted to the Copernican system by its greater simplicity and economy relative to the Ptolemaic system. He believed that it was in his power to establish the heliocentric view beyond any shadow of a doubt by doing the one thing that no one else had previously done successfully; namely, to discover an angle of parallax for a fixed star. To this end he designed and built his perpendicular telescope. Although his results were unsatisfactory (the angle he claimed to find was much greater than anything even remotely possible) according to modern standards, his insistence upon the truth of the Copernican view and his efforts to defend it scientifically served as a stimulus to his contemporaries to do likewise. It is quite possible that he could have done otherwise (as did Brahe) and thereby hindered the future development of the heliocentric theory.

The reason for Hooke's faulty measurements (his conclusion, after all, was correct) was twofold. For one thing, he was working with very crude instruments. For another, his strong belief in the truth of the conclusion to be established made him overlook the shortcomings of his instruments and seize upon any bit of available data as confirming the Copernican theory. Hooke speaks in his first Cutler Lecture, concerning the proof for the annual motion of the earth, as the man the world has been waiting for; the scientific messiah who would put to rout the numerous adversaries of truth. Understandably, such zealousness made the Curator's eyes and lenses better than they actually were.

Similar observations can be put forward with respect to the Curator's interest in the daily motion of the earth. According to Hooke, and again, as in the case of the path taken by a freely falling body as it approached the earth's center, he does not tell us exactly why he thought so, a body dropped from a height above a moving earth should fall to the S. S. E. of the point of release. This he attempted to demonstrate by means of an experiment which he repeated several times. Again, although his results were not satisfactory based upon modern calculations, his widely publicized work tended to perpetuate interest in the topics and belief in the truth of the proposition that the earth, and not the heavens, turns a circle in twenty-four hours.

Hooke also believed he had something worthwhile to say concerning the reason why the planets continue to traverse an elliptical path around the sun. Using a pendulum to illustrate his thesis, he maintained that a curved path around a center was due to a combination of two motions: one tending to move a body in a straight line tangent and away from the center and another moving the body toward the center. It was shown that he could not have gotten this information from either Descartes or Borelli.

Hooke was also very much interested in the cause of the "force" that pulled the planets toward the sun; namely, gravitation. This he believed to be exactly the same in nature as the cause of heavy bodies near the surface of the earth moving toward the earth's center. In fact, every body attracted every other body in the universe, a capital advance in thought. In his own mind, Hooke, after rejecting some form of explanation based upon magnetism, resolved this issue by postulating millions of vibrations per second in the subtle material ether which tended to carry the bodies in question "down" toward the center of vibration. This power of gravitation, thought the Curator, diminished as one became farther removed from the center of vibration due to the slowing down and diffusion of the etherial vibrations. One sees in gravitation another example of mechanism at work in nature.

But, exactly how much does the power of gravity diminish at any particular distance? The decrease in pull, claimed Hooke, was inversely proportional to the square of the distance. This is explicitly stated in his correspondence with Newton previous to the writing of the *Principia*. Hooke could not or did not, however, as far as we know, prove his thesis mathematically; Newton did.

Nevertheless, Hooke should not be regarded as in all ways inferior to Newton. Before we can say something meaningful on the relationships between Hooke, Newton, and modern science, we must turn our attention to the distinction between a philosophical and a mathematical approach to nature. If we define the philosophy of nature as the attempt to penetrate the most basic and comprehensive causes (causes here is understood as meaning one or more of the four Aristotelian types) underlying observed natural phenomena, and if we define the mathematical approach to nature (which I will also call mathematical physics) as the attempt to develop a purely formal system of explanation which deliberately prescinds from any consideration of the real, physical causes underlying the phenomena which are mathematically treated, we can say that Hooke was primarily interested in the former while Newton was primarily interested in developing and expounding upon the latter. This distinction is not absolute in the sense that if a person attempts one he cannot also attempt the other. There is no necessity that they mutually exclude each other in the thinking of any one individual. It is highly unusual, though, for one person to emphasize both ways of looking at nature. This difference in emphasis is highly important when comparing Hooke's efforts in the science of mechanics with the effort of Newton. Hooke was primarily a philosopher of nature while Newton was *primarily* a mathematician of nature, if we may so speak. Once this difference is established and clearly recognized there can be no talk of Newton as a success and Hooke as a failure at the game of science for it will be seen that they were not playing the same game.

That Newton was primarily interested in giving a mathematical explanation for the phenomena of gravitation and planetary motion can be seen from the very title, and also from his own original Preface, to his major work. We read as the very first lines in his *Mathematical Principles of Natural Philosophy* that: "Since the ancients (as we are told by Pappus) esteemed the science of mechanics of greatest importance in the investigation of natural things, and the moderns, rejecting substantial forms and occult qualities, have endeavored to subject the phenomena of nature to the laws of mathematics, I have in this treatise cultivated mathematics as far as it relates to philosophy."

This does not mean that Newton scorned experiments and observations. For, indeed, farther down in the same Preface we hear him say that "the whole burden of philosophy seems to consist in this – from the phenomena of motions to investigate the forces of nature, and then from these forces to demonstrate the other phenomena." This agrees perfectly with what Newton states concerning the complementarity of the method of synthesis or composition (i.e., the deduction of conclusions from principles) and the method of analysis (i.e., the induction of principles from facts) in book III, query 31 of his *Opticks*. The whole emphasis of Newton's endeavors, however, was not on experimentation and the search after the true, physical causes of natural occurrences but rather on mathematics, the purely formal and hollow description of events. He well knew that a mathematical formula was quite different from a causal explanation.

This distinction was not a part of one's general educational background in the seventeenth century. Hooke never explicitly recognized any such distinction. At times, Hooke appears highly mathematical, as, for example, when he is carefully dropping weights and carefully noting measurements in order to test a hypothesis. At other times, as, for instance, when discussing the existence and properties of the ether, he seems inexorably bound to philosophy. At all times, however, it was Hooke's explicit aim and desire to know nature as it really is; to penetrate to the heart of its mysteries and reveal to his fellow man the wonders of creation and its

Creator. As Hooke himself affirms near the beginning of his "A General Scheme,"

> I do not here with the Scepticks affirm, that nothing is or can be known, my Design is quite another thing; their end only in denying any thing to be knowable, seems to be Dispute, and tends to Ignorance and Laziness, mine on the other side supposes all things as possible to be known, and accordingly studies and considers of the Means that seem to tend to that end.

For Hooke there were no truly "occult qualities" in nature. If some things were hidden in the seventeenth century it was only a temporary state of affairs. In time the veil of ignorance would be lifted and man would know all the secrets of nature. It is because of this over-riding *Weltanschauung* that one can firmly assert that Hooke was more the philosopher of nature than its mathematical investigator. Within the framework of Cartesian physics and the Baconian induction, Hooke hoped one day to know in detail nature as it really is in itself. This knowledge would then benefit mankind.

Rather than comparing Hooke to Newton, it would perhaps be more fair to compare Hooke to someone like Thomas Edison or Alexander Graham Bell. Edison, for example, was a man who slept little, experimented constantly, and whose work resulted in several devices which have greatly increased the physical welfare of mankind. According to the Newtonian-type standard of success, Edison would also be a failure. One can see here the relativity of the terms success and failure, even in the scientific sphere. Let us not attempt to apply the same yardstick to everyone, even in the sciences.

In fine, then, what can we say concerning Robert Hooke's position in seventeenth century natural philosophy, with special emphasis upon his mechanics? Our conclusion is that Hooke represents the last great Baconian in the history of science. Others after Hooke attempted to interpret the achievements of science in terms of a more or less exclusively inductive process, but Hooke was the last to actually attempt to achieve results in natural philosophy by adhering to the Baconian emphasis upon induction. The new approach of mathematical physics, which did not depend upon more or less total and complete inductions, ushered in by Newton, even during Hooke's own lifetime, was largely foreign to the Curator. Hooke was interested in gathering facts, delving into the real causes of natural phenomena, and using his knowledge and talents in order to make life on earth easier for mankind. The result of this attitude was a long string of brilliantly conceived mechanical devices of all sorts. In mechanics, however, the Curator has been eclipsed by Kepler, Galileo, Huygens and Newton. Yet, it seems clear to this author that, if Hooke were given the choice of being remembered either for something such as the ordinary matchstick or pocket watch of today or for the whole Newtonian synthesis, he would have preferred to be the inventor of some practical device which has so greatly added to man's comfort and well-being.

# BIBLIOGRAPHY

*1. Primary Sources*

A. Books


*A discourse of a New Instrument to make more accurate observations in Astronomy, than ever were yet made,* London, 1661.

*An Attempt for the Explication of the Phenomena Observable in an Experiment Published by the Honourable Robert Boyle,* London, 1661.

*Reponse de Monsieur Hook aux considerations de M. Auzout,* Paris, 1665.

*Micrographia: or some Physiological Descriptions of Minute Bodies made by Magnifying Glasses with Observations and Inquiries thereupon,* London, 1665.

*An Attempt to Prove the Motion of the Earth from Observations made by Robert Hooke Fellow of the Royal Society,* London, 1674.

*Animadversions On the first part of the Machina Coelestis Of the Honourable, Learned, and deservedly Famous Astronomer Johannes Hevelius Consul of Dantzick,* London, 1674.

*A Description of Helioscopes, And some other Instruments made by Robert Hooke, Fellow of the Royal Society*, London, 1676.

*Lampas: or, Descriptions of some Mechanical Improvements of Lamps and Waterpoises. Together with some other Physical and Mechanical Discoveries,* London, 1677.

*Lectures and Collections Made by Robert Hooke, Secretary of the Royal Society: Cometa and Microscopium,* London, 1678.

*Lectures De Potentia Restitutiva, or of Spring Explaining the Power of Springing Bodies,* London, 1678.

*Lectiones Cutlerianae, or a Collection of Lectures: Physical, Mechanical, Geographical, and Astronomical,* London, 1679. (This work is a collection of the six foregoing treatises.)

*The Posthumous Works of Robert Hooke, M.D. S.R.S. Geom. Prof. Gresh. Etc.* (ed. by Richard Waller), London, 1705.

*The Diary of Robert Hooke, 1672-1680* (ed. by H. W. Robinson and W. Adams), London, 1935.

"The Diary of Robert Hooke: Part I: 1 November 1688 to 9 March 1690; Part II: 6 December 1692 to 8 August 1693," *Early Science in Oxford* (ed. by R. W. T. Gunther), Vol. 10, Oxford, England, 1935, pp. 69-265.

B. Articles

"Introduction," *The English Atlas* by Moses Pitt, Vol. I, London, 1680.

"Preface," *An Historical Relation of the Island Ceylon, in the East-Indies* by Robert Knox, London, 1681.

"An Optical Discourse by R. H. proposing a way of helping short-sighted or purblind Eyes," *Philosophical Collections*, No. 3 (December 1681), pp. 59-60.

"A Mechanical Discourse Containing the Description of the best form of Horizontal Sayls for a Mill, and the ground of the inclined Sayls of Ships by R.H.," *Ibid.*, pp. 61-64.

The following articles all appeared in the *Philosophical Transactions,* Vols. 1-17.

"A Spot in one of the belts of Jupiter," Vol. I (1665-1666), p. 3.

"Mr. Hook's Answer to Monsieur Auzout's Considerations, in a Letter to the Publisher of these Transactions," *Ibid.*, pp. 63-73.

"A Method, by which a Glass of a small Plano-convex Sphere may be made to refract the Rayes of light to a Focus of a far greater distance, than is usual," *Ibid.*, pp. 202-203.

"A new Contrivance of Wheel-Barometer, much more easy to be prepared, than that, which is described in the *Micrography;* imparted by the Author of that Book," *Ibid.*, pp. 218-219.

"The Particulars Of those Observations of the Planet Mars, formerly intimated to have been made at London in the Months of February and March A. 1665/6," *Ibid.*, pp. 239-242.

"Some Observations Lately made at London concerning the Planet Jupiter and A late Observation about Saturn made by the same," *Ibid.*, pp. 245-247.

"Observations made in several places, Of the late Eclipse of the Sun, which hapned on the 22 of June, 1666," *Ibid.*, pp. 295-296.

"Directions for Observations and Experiments to be made by Masters of Ships, Pilots, and other fit Persons in their Sea-Voyages," Vol. 2 (1667), pp. 433-448.

"More Wayes For the same Purpose of dividing a foot into many thousand parts, Intimated by M. Hook," *Ibid.*, p. 459.

"An Account of an Experiment made by Mr. Hook, of Preserving Animals alive by Blowing through their Lungs with Bellows," *Ibid.*, pp. 539-540.

"A Description Of an Instrument for dividing a foot into many thousand parts," *Ibid.*, pp. 541-544.

"A Contrivance To make the Picture of any thing appear on a Wall, Cub-board, or within a Picture-frame, etc., in the midst of a Light room in the day-time; or in the Night-time in any room that is enlightened with a considerable number of Candles," Vol. 3 (1668), pp. 741-743.

"Some Communications, confirming the present Appearances of the Ring about Saturn, by M. Hugens de Zulechem and Mr. Hook," Vol. 5 (1670), p. 2093.

"Observations made by Mr. Hook of some Spots in the Sun August 30, 1671 and Observations made by the same of the late Eclipse of the Moon, the 8th. September, 1671," Vol. 6 (1671), pp. 2295-2296.

"An Account of what hath been observed here in London and Derby, by Mr. Hook, Mr. Flamsteed, and others, concerning the late Eclipse of the Moon of Jan. 1, 1674/5," Vol. 9 (1674), p. 237.

"Two Astronomical Observations of the Eclipses of the Planet Jupiter, by the Moon in March and April last, made at London," Vol. 16 (1688), pp. 85-86.

"A Description of an Invention, whereby the Divisions of the Barometer may be enlarged in any given proportion," *Ibid.*, pp. 241-244.

"A Relation of the great effects of a new sort of Burning Speculum lately made in Germany: taken from the *Acta Eruditorum* of the Month of January last," *Ibid.*, pp. 352-354.

"The Abstract of two Letters sent some time since by Mr. Anth. Van Leeuwenhoeck to Dr Gale and Dr Hooke," Vol. 17 (1693), pp. 593-594.

"Additional Note to a paper by Dr. Lister giving a transcript from *Micrographia* on the wings of insects," *Ibid.*, pp. 691-692.

C. Hooke *in aliis*

Sprat, Thomas, *The History of the Royal-Society of London, For the Improving of Natural Knowledge*, London, 1667.

Halley, Edmund, "An Account of Dr Robert Hook's Invention of the Marine Barometer, with its Description and Uses by E. Halley, R.S.S.," *Philosophical Transactions*, Vol. 22 (1702), pp. 791-794.

Halley, Edmund (ed.), *Miscellanea Curiosa Being a Collection Of some of the Principal Phaenomena in Nature*, London, 1705.

Derham, William, *Philosophical Experiments and Observations of the late eminent Dr. Robert Hooke, S.R.S. and geom. prof. Gresh., and other eminent virtuoso's in his time*, London, 1726.

Birch, Thomas, *The History of the Royal Society of London*, 4 vols., London, 1756-1757.

Gunther, R. W. T., *Early Science in Oxford*, Vols. 6, 7, Oxford, England, 1930.

2. *Secondary Sources*

Andrade, Edward, "Robert Hooke and His Contemporaries," *Nature*, Vol. 136 (1935), pp. 358-361.

—, "The Real Character of Bishop Wilkins," *Annals of Science*, Vol. I (1936), pp. 4-12.

—, "Robert Hooke," *Proceedings of The Royal Society*, Series B, Vol. 137 (1950), pp. 153-186.

—, "Robert Hooke, Inventive Genius," *The Listener*, 8 February 1951.

—, "Robert Hooke, 1635-1703," *Nature*, Vol. 71 (1953), pp. 365-367.

—, *Sir Isaac Newton*, Garden City, New York, 1958.

—, *A Brief History of the Royal Society*, London, 1960.

Armitage, Angus, "The Deviation of Falling Bodies," *Annals of Science*, Vol. 5 (1941-1947), pp. 342-351.

—, "Borell's Hypothesis and the Rise of Celestial Mechanics," *Annals of Science*, Vol. 6 (1948-1950), pp. 268-282.

—, *Edmond Halley*, Edinburgh, 1966.

Ashby, Eric, *Technology and the Academics: An Essay on Universities and the Scientific Revolution*, New York, 1958.

Aubrey, John, *Brief Lives Chiefly of Contemporaries, set down by John*

*Aubrey, between the years 1669 and 1696* (ed. by Andrew Clark), 2 vols., Oxford, 1898.

—, *Aubrey's Brief Lives* (ed. by O. L. Dick), Ann Arbor, Michigan, 1957.

Auzout, M., "Considerations upon Mr. Hook's New Instrument for Grinding of Optick-Glasses," *Philosophical Transactions,* Vol. I (1665-1666), pp. 56-62.

Bacon, Francis, *Sylva Sylvarum, Natural History of Life and Death, Articles of Enquiry Touching Metals and Minerals, New Atlantis* (ed. by W. Rawley), 9th ed., London, 1669.

—, *The Advancement of Learning and Novum Organum* (tr. and ed. by J. E. Creighton), Revised ed., New York, 1900.

Bannerman, W. Bruce (ed.), *The Registers of St. Helen's,* London, 1904.

Bartholine, Thomas, *Acta Hafniensia,* Copenhagen, 1673.

Bell, Arthur E., *Christiaan Huygens and the Development of Science in the Seventeenth Century,* New York, 1947.

Bell, Walter G., *The Great Fire of London in 1666,* London, 1920.

—, *The Great Plague in London in 1665,* London, 1924.

Blackwell, Richard J., "Descartes' Laws of Motion," *Isis,* Vol. 57 (1966), pp. 220-234.

Boas, Marie, *Robert Boyle and Seventeenth-Century Chemistry,* Cambridge, England, 1958.

Borelli, Giovanni, *Theoricae Mediceorum Planetarum ex causis physicis deductae,* Florentiae, 1666.

—, *Risposta di Gio,* Messina, 1668.

Boyle, Robert, *The Works of the Honourable Robert Boyle* (ed. by Thomas Birch), 6 vols., London, 1772.

—, *Robert Boyle on Natural Philosophy* (ed. by M. B. Hall), Bloomington, Indiana, 1965.

Brewster, David, *Memoirs of the Life, Writings, and Discoveries of Sir Isaac Newton,* 2 vols., Edinburgh, 1855.

Burtt, Edwin A., *The Metaphysical Foundations of Modern Physical Science,* London, 1925.

Butterfield, Herbert, *The Origins of Modern Science,* New York, 1962.

Cajori, Florian, "Newton's Twenty Years' Delay in Announcing the Law of Gravitation," *Sir Isaac Newton 1727-1927* (ed. by the History of Science Society), Baltimore, 1928, pp. 127-188.

—, *A History of Physics,* New York, 1929.

Centore, F. F., "Copernicus, Hooke and Simplicity," *Philosophical Studies,* Vol. 17 (1968), pp. 185-196.

Chalmers, Alexander, "Robert Hooke," *The General Biographical Dictionary,* New ed., Vol. 18, London, 1814, pp. 128-135.

Chancellor, E. Beresford, *Lives of the British Architects,* London, 1909.

Clagett, Marshall, *Archimedes in the Middle Ages,* Madison, Wisconsin, 1964.

Clark, Andrew, *The Life and Times of Anthony Wood,* 5 vols., Oxford, 1891-1900.

Clave, Etienne de, *Nouvelle Lumiere Philosophique,* Paris, 1641.

Conant, James B., *On Understanding Science: An Historical Approach,* New Haven, Conn., 1947.

—, "Robert Boyle's Experiments in Pneumatics," *Harvard Case Histories in*

*Experimental Science* (ed. by J. B. Conant and L. K. Nash), Vol. I, Cambridge, Mass., 1957, pp. 3-63.
Crowther, James G., *Founders of British Science; Wilkins, Boyle, Ray, Wren, Hooke, Newton*, London, 1960.
Cummings, Alexander, *Elements of Clock-Work*, London, 1766.
Derham, William, *The Artificial Clock-Maker*, 3rd ed., London, 1714.
Descartes, René, *René Descartes: Philosophical Writings* (tr. and ed. by Elizabeth Anscombe and Peter T. Geach), Edinburgh, 1964.
*The Dictionary of National Biography* (ed. by L. Stephen and S. Lee), 22 vols., Oxford, England, 1917.
Dijksterhuis, Eduard J., *The Mechanization of the World Picture* (tr. by C. Dikshoorn), Oxford, England, 1961.
— and Forbes, R. J., *A History of Science and Technology*, 2 vols., Baltimore, 1963.
Dugas, Rene, *The History of Mechanics* (tr. by J. R. Maddox), London, 1957.
—, *Mechanics in the Seventeenth Century* (tr. by F. Jacquot), Neuchatel, Switzerland, 1958.
Elmes, James, *Sir Christopher Wren and His Times*, London, 1852.
'Espinasse, Margaret, *Robert Hooke*, London, 1956.
Farrington, Benjamin, *Francis Bacon, Philosopher of Industrial Science*, New York, 1949.
Freeman, Edmund, "A Proposal for an English Academy in 1660," *Modern Language Review*, Vol. 19 (1924), pp. 291-300.
Fulton, John F., *A Bibliography of the Honourable Robert Boyle*, 2nd ed., Oxford, England, 1961.
Galilei, Galileo, *Dialogues Concerning Two New Sciences* (tr. by H. Crew and A. de Salvio), New York, 1914.
—, *Dialogue on the Great World Systems* (tr. by T. Salusbury, ed. by G. de Santillana), Chicago, 1953.
Grant, Robert, *The History of Physical Astronomy*, London, 1852.
Gunther, R. W. T., *Historic Instruments for the Advancement of Science*, London, 1925.
—, *Early Science in Oxford*, 14 vols., Oxford, England, 1920-1945.
Haldane, Elizabeth Sanderson, *Descartes: His Life and Times*, New York, 1905.
Hall, A. Rupert, "Two Unpublished Lectures of Robert Hooke," *Isis*, Vol. 42 (1951), pp. 219-230.
—, "Robert Hooke and Horology," *Notes and Records of the Royal Society*, Vol. 8 (1951), pp. 167-177.
—, *The Scientific Revolution, 1500-1800*, London, 1954.
—, "Newton on the Calculation of Central Forces," *Annals of Science*, Vol. 13 (1957), pp. 62-71.
—, *From Galileo to Newton, 1630-1720*, London, 1963.
— and Marie B., "Why Blame Oldenburg?" *Isis*, Vol. 53 (1962), pp. 482-491.
— (eds.), *The Correspondence of Henry Oldenburg*, 6 vols., Madison, Wisconsin, 1965-1969.
Hartley, Harold B., *The Royal Society: Its Origins and Founders*, London, 1960.

Hartmann, Philippus J., *Succini Prussici Historia Physica et Civilis,* Francofurti, 1677.

Hearne, Thomas, *Reliquiae Hearnianae* (ed. by P. Bliss), 2 vols., London, 1869.

Herivel, John, *The Background to Newton's Principia,* Oxford, England, 1965.

Hesse, M. B., "Hooke's Philosophical Algebra," *Isis,* Vol. 57 (1966), pp. 67-83.

—, *Forces and Fields,* New York, 1961.

Horrox, Jeremiah, *Opera Posthuma* (ed. by J. Wallis), London, 1673.

Houghton, W. E., Jr., "The History of Trades: Its Relation to Seventeenth Century Thought," *The Roots of Scientific Thought* (ed. by P. P. Wiener and A. Noland), New York, 1957, pp. 354-381.

Huygens, Christiaan, *Treatise on Light* (tr. by S. P. Thompson), Chicago, 1912.

—, *Horologium Oscillatorium,* Paris, 1673.

—, *Opuscula Posthuma,* Amsterdam, 1703.

James, E. Bouchar, *Letters Archeological and Historical Relating to the Isle of Wight,* Vol. 2, London, 1896.

James, W. S., "The Discovery of the Gas Laws," *Science Progress,* Vol. 23 (1928), pp. 263-272.

Jammer, Max, *Concepts of Force,* New York, 1962.

Johnson, F. R., "Gresham College: Precursor of the Royal Society," *The Roots of Scientific Thought* (ed. by P. P. Wiener and A. Noland), New York, 1957, pp. 328-353.

Jones, Richard F., *Ancients and Moderns: A Study of the Rise of the Scientific Movement in Seventeenth Century England,* 2nd ed., St. Louis, 1961.

Jourdain, Philip E. B., "Robert Hooke as a Precursor of Newton," *Monist,* Vol. 23 (1913), pp 353-384.

Kent, William, *Encyclopedia of London,* London, 1937.

Keynes, Geoffrey, *A Bibliography of Dr. Robert Hooke,* Oxford, England, 1960.

Koyré, Alexander, "A Note on Robert Hooke," *Isis,* Vol. 41 (1950), pp. 195-196.

—, "An Unpublished Letter to Isaac Newton (December 9, 1679)," *Isis,* Vol. 43 (1952), pp. 312-337.

—, "A Documentary History of Fall from Kepler to Newton," *Transactions of the American Philosophical Society,* Vol. 45 (1955), part 4, pp. 329-395.

—, *La Révolution Astronomique,* Paris, 1961.

—, *Newtonian Studies,* Cambridge, Mass., 1965.

Linus (Hall), Francis, *Tractatus de Corporum Inseparabilitate,* London, 1661.

Lohne, Johannes, "Hooke versus Newton," *Centaurus,* Vol. 7 (1960), pp. 6-52.

Lyons, Henry George, *The Royal Society, 1660-1940: A History of Its Administration Under Its Charters,* Cambridge, England, 1944.

Mach, Ernst, *The Science of Mechanics* (tr. by T. J. McCormack), 9th ed., La Salle, Ill., 1942.

Mackay, Andrew, *The Complete Navigator: or An Easy and Familiar Guide to the Theory and Practice of Navigation,* 2nd ed., London, 1810.

Maddison, R. E. W., "Studies in the Life of Robert Boyle," *Notes and Records of the Royal Society,* Vol. 9 (1951), pp. 1-35.

Magalotti, Lorenzo, *Essayes of Natural Experiments made in the Academie del Cimento* (tr. by R. Waller), London, 1684.

Martin, Benjamin, *Biographia Philosophica*, London, 1764.
Masson, Flora, *Robert Boyle: A Biography*, London, 1914.
Masson, Irvine, "The Genesis of the Royal Society," *Nature*, Vol. 113 (1924), pp. 197-199.
McKie, Douglas, "Fire and the Flamma Vitalis: Boyle, Hooke and Mayow," *Science, Medicine and History* (ed. by E. A. Underwood), Vol. I, London, 1953, pp. 469-488.
McRae, Robert, *The Problem of the Unity of the Sciences: Bacon to Kant*, Toronto, 1961.
Merton, Robert K., "Science, Technology and Society in Seventeenth Century England," *Osiris*, Vol. 4 (1938), pp. 360-632.
Meyerson, Emile, *Identity and Reality* (tr. by K. Loewenberg), London, 1930.
Millington, E. C., "Theories of Cohesion in the Seventeenth Century," *Annals of Science*, Vol. 5 (1941-1947), pp. 253-269.
Mohler, Nora, "The Spring and Weight of the Air," *The American Physics Teacher*, Vol. 7 (1939), pp. 380-389.
Montucla, Jean E., *Histoire des Mathématiques*, Vol. 4, Paris, 1802.
More, Louis T., *The Life and Works of the Honourable Robert Boyle*, London, 1944.
—, *Isaac Newton*, New York, 1962.
Newton, Isaac, *Mathematical Principles of Natural Philosophy* (tr. by Andrew Motte; revised and ed. by Florian Cajori), Berkeley, California, 1947.
—, *Opticks*, 4th ed., London, 1730.
Olmsted, J. W., "The 'Application' of Telescopes to Astronomical Instruments 1667-1669: A Study in Historical Method," *Isis*, Vol. 40 (1949), pp. 213-225.
Ornstein-Bronfenbrenner, Martha, *The Role of the Scientific Societies in the Seventeenth Century*, New York, 1913.
Pascal, Blaise, *The Physical Treatises of Pascal. The Equilibrium of Liquids and the Weight of the Mass of the Air* (tr. by I. H. B. and A. G. H. Spiers), New York, 1937.
Patterson, Louise D., "Robert Hooke and the Conservation of Energy," *Isis*, Vol. 38 (1948), pp. 151-156.
—, "Hooke's Gravitation Theory and its Influence on Newton," *Isis*, Vol. 40 (1949), pp. 327-341; Vol. 41 (1950), pp. 32-45.
—, "A Reply to Professor Koyré's Note on Robert Hooke," *Isis*, Vol. 41 (1950), pp. 304-305.
—, "Pendulums of Wren and Hooke," *Osiris*, Vol. 10 (1952), pp. 279-321.
Patterson, T. S., "John Mayow in Contemporary Setting," *Isis*, Vol. 15 (1931), pp. 47-96; 504-546.
Pelseneer, Jean, "Une Lettre Inédite de Newton (December 13, 1679)," *Isis*, Vol. 12 (1929), pp. 237-254.
Petty, William, *Duplicate Proportion together with a New Hypothesis of Springing or Elastique Motions*, London, 1674.
Picard, Jean, *Mesure de la Terre*, Paris, 1671.
Pledge, Humphrey T., *Science Since 1500*, New York, 1959.
Powell, Anthony, *John Aubrey and his Friends*, London, 1948.
Power, Henry, *Experimental Philosophy, in Three Books: Containing New Experiments Microscopical, Mercurial, Magnetical*, London, 1664.

Purver, M. and Bowen, E. J., *The Beginning of the Royal Society,* Oxford, England, 1960.

Purver, Margery, *The Royal Society: Concept and Creation,* Cambridge, Mass., 1967

Reilly, Conor, "Francis Line, Peripatetic (1595-1675)," *Osiris,* Vol. 14 (1962), pp. 222-253.

Rix, Herbert, "Henry Oldenburg, First Secretary of the Royal Society," *Nature,* Vol. 49 (1893-1894), pp. 9-12.

Robison, John, "Robert Hooke," *Encyclopaedia Britannica,* 3rd ed., Vol. 8, Edinburgh, 1797, pp. 645-646.

—, "Robert Hooke," *Encyclopaedia Britannica,* Supplement to 3rd ed., Vol. I, Edinburgh, 1801, pp. 742-743.

Royal Society (ed.), *The Record of the Royal Society of London,* 3rd ed., London, 1912.

—, *The signatures in the first Journal-book and the Charter-book of the Royal Society, being a facsimile of the signatures of the founders, patrons and fellows of the society from the year 1660 down to the present time,* London, 1912.

Sabra, A. I., *Theories of Light From Descartes To Newton,* London, 1967.

Sargeaunt, John, *Annals of Westminster School,* London, 1898.

Sarton, George, "Simon Stevin of Bruges," *Isis,* Vol. 21 (1934), pp. 241-303.

—, *The Appreciation of Ancient and Medieval Science During the Renaissance,* New York, 1961.

Smith, John, *Horological Disquisitions,* London, 1694.

Stimson, Dorothy, *Scientists and Amateurs, a History of the Royal Society,* New York, 1948.

Syfret, R. H., "The Origins of the Royal Society," *Notes and Records of the Royal Society,* Vol. 5 (1948), pp. 75-137.

Symonds, Robert Wemyss, *Thomas Tompion,* London, 1951.

Tait, Peter G., "Hooke's Anticipation of the Kinetic Theory and Synchronism," *Proceedings of the Royal Society of Edinburgh,* Vol. 13 (Nov. 1884-July 1886), p. 118.

Taylor, E. G. R., *Mathematical Practitioners of Tudor and Stuart England,* Cambridge, England, 1954.

Thomson, Thomas, *The History of the Royal Society,* London, 1812.

Thorndike, Lynn, "Newness and Novelty in Seventeenth-Century Science and Medicine," *The Roots of Scientific Thought* (ed. by P. P. Wiener and A. Noland), New York, 1957, pp. 443-457.

Turnbull, Herbert W., *James Gregory,* London, 1939.

— and Scott, J. F. (eds.), *The Correspondence of Isaac Newton,* 4 vols., Cambridge, England, 1959-1967.

Wallis, John, *A Defence of the Royal Society: An Answer to the Cavils of Dr. William Holder,* Oxford, 1678.

—, "A Summary Account given by Dr. John Wallis of the General Laws of Motion," *Philosophical Transactions,* Vol. 3 (1668), pp. 864-866.

Ward, Edward, *A Compleat and Humorous Account of all the Remarkable Clubs and Societies in the Cities of London and Westminster,* 7th ed., London, 1756.

Ward, John, *Lives of the Professors of Gresham College,* London, 1740.
Weld, Charles R., *The History of the Royal Society,* 2 vols., London, 1848.
Westfall, R. S., "Newton's Reply to Hooke and the Theory of Colors," *Isis,* Vol. 54 (1963), pp. 82-96.
Whewell, William, *History of the Inductive Sciences,* 3 vols., London, 1837.
Whiston, William, *Memoirs of the Life of Mr. William Whiston by Himself,* 2 vols., London, 1749-1750.
Whittaker, Edmund T., *A History of the Theories of Aether and Electricity,* Vol. I, Edinburgh, 1951.
Wilchinsky, Z., "The Theoretical Treatment of Hooke's Law," *The American Physics Teacher,* Vol. 7 (1939), p. 134.
Wilkins, John, *Mathematical Magick,* London, 1634.
Williams, E., "Hooke's Law and the Concept of the Elastic Limit," *Annals of Science,* Vol. 12 (1956), pp. 74-83.
Wolf, Abraham, *A History of Science, Technology and Philosophy in the Sixteenth and Seventeenth Centuries,* 2nd ed., 2 vols., New York, 1959.
Wood, Anthony, *The History and Antiquities of the University of Oxford,* Oxford, England, 1674.
—, *Athenae Oxonienses,* 2nd ed., Vol. 2, London, 1721, cols. 1039-1041.
Young, Thomas, *A Course of Lectures on Natural Philosophy,* Vol. 1, London, 1807.

## INDEX TO SELECTED PROPER NAMES

GPSR Compliance

*The European Union's (EU) General Product Safety Regulation (GPSR) is a set of rules that requires consumer products to be safe and our obligations to ensure this.*

*If you have any concerns about our products, you can contact us on ProductSafety@springernature.com*

In case Publisher is established outside the EU, the EU authorized representative is:

Springer Nature Customer Service Center GmbH
Europaplatz 3
69115 Heidelberg, Germany

**Batch number: 10607224**

Printed by Printforce, the Netherlands